BEAUTY OVER TROUBLED WATER

THE BRIDGES OF NIAGARA FALLS

RICHARD PANCHYK

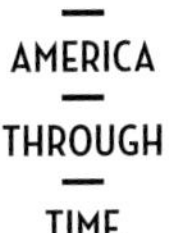

AMERICA
THROUGH
TIME

America Through Time
Fonthill Media LLC
www.through-time.com
books@through-time.com

First published 2025

ISBN 978-1-62545-148-4

Typeset in Mrs Eaves XL Serif Narrow
Printed and bound in England

Acknowledgments

Thanks to H&H for the use of their extensive archive of incredible photographs documenting the construction of the Rainbow Bridge, Lewiston-Queenston Bridge, and North and South Grand Island Bridges. Thanks to Kena Smith for her help with the editorial process.

All images courtesy of the author unless otherwise stated.

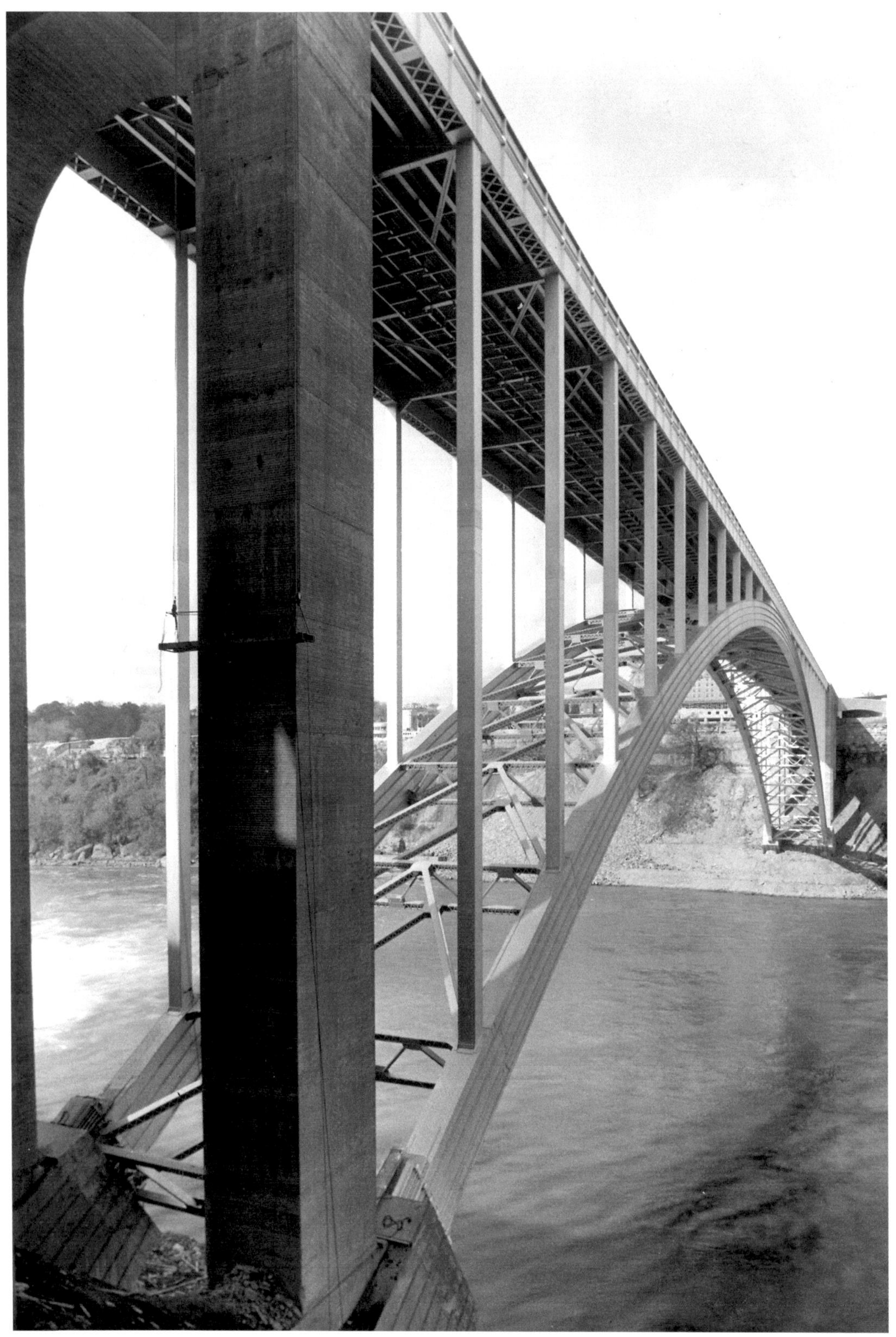

PREFACE

As the unofficial historian of H&H (Hardesty & Hanover), where I have worked since 2012, I have long been fascinated by one of the company's most well-known structures—the Rainbow Bridge at Niagara Falls. The company's history goes all the way back to 1887, but the Rainbow Bridge stands as one of their hallmark achievements. The firm has an extensive photographic archive of hundreds of images documenting the construction and opening of the Rainbow Bridge, as well as their "sequel" bridge, the Lewiston-Queenston Bridge. Those photographs were, in part, the inspiration for this book.

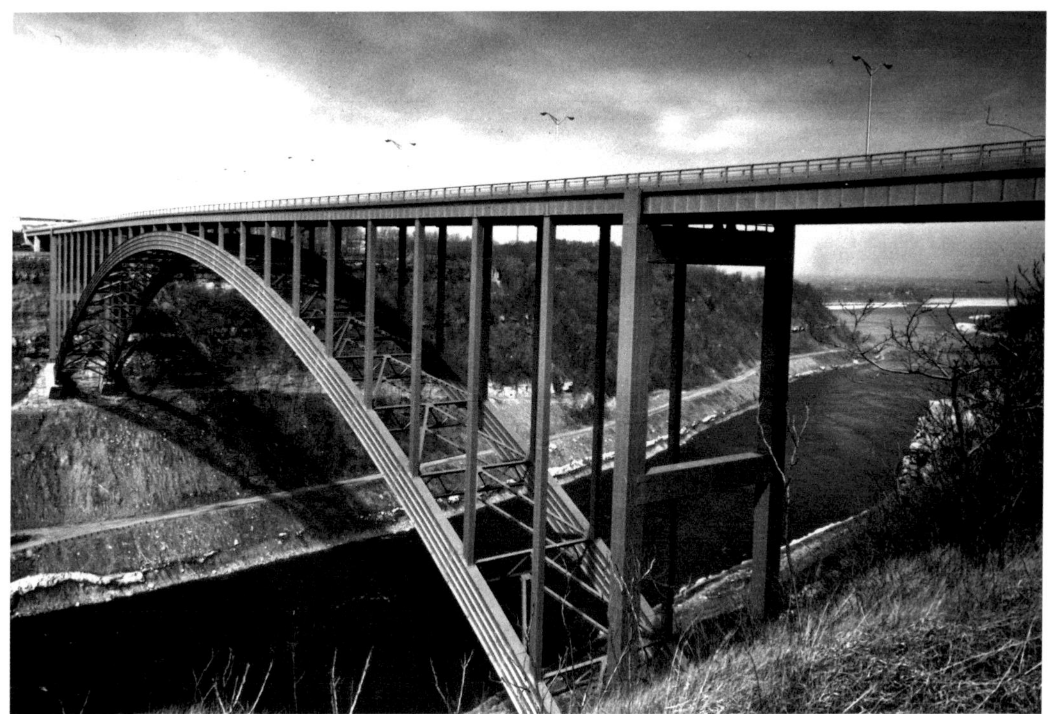

Introduction

Extending just 36 miles from Lake Erie to Lake Ontario, the Niagara River is one of more than 250,000 rivers in the United States. By length alone it is certainly not remarkable at all (the longest river in the country, the Missouri, is 2,341 miles in length), but it has gained international fame because it happens to host one of nature's most spectacular wonders—the world-famous Niagara Falls. There are more than 7,800 documented waterfalls around the world, but perhaps none are so impressively dramatic, accessible, extensive, photogenic, and tourist friendly as Niagara Falls. Adding to its appeal, the Niagara River and its falls also form part of the boundary between Canada and the United States, making it an added attraction as a border crossing point for those who want to visit our northerly neighbor. Great views of the falls can be had from both sides of the border. Niagara Falls is also one of the top honeymoon destinations in North America.

What starts out as a calm and relatively ordinary river less than 20 miles south between Buffalo, New York, and Fort Erie, Ontario, Canada, turns into a roiling, raging, beautiful scene ringed by the several waterfalls of which Niagara Falls is comprised—the Horseshoe Falls, the American Falls, and the Bridal Veil Falls. These waterfalls were formed fairly recently, in the overall geological timeline, as a product of the last ice age about 10,000–12,000 years ago, as water from melting glaciers eroded rock layers and tumbled over steep cliffs.

The area was first seen by Europeans in the seventeenth century, and settlers to the New World were fascinated by this wondrous place. Given the growing popularity of the falls as a tourist destination and the fact that they straddled America and Canada, there was a necessity to build bridges across the river. That effort began in the mid-nineteenth century once technology allowed it. Many of those first structures were sooner or later destroyed by the elements (especially wind and snow/ice) until more recent iterations were better able to withstand the forces of nature thanks to much improved technology.

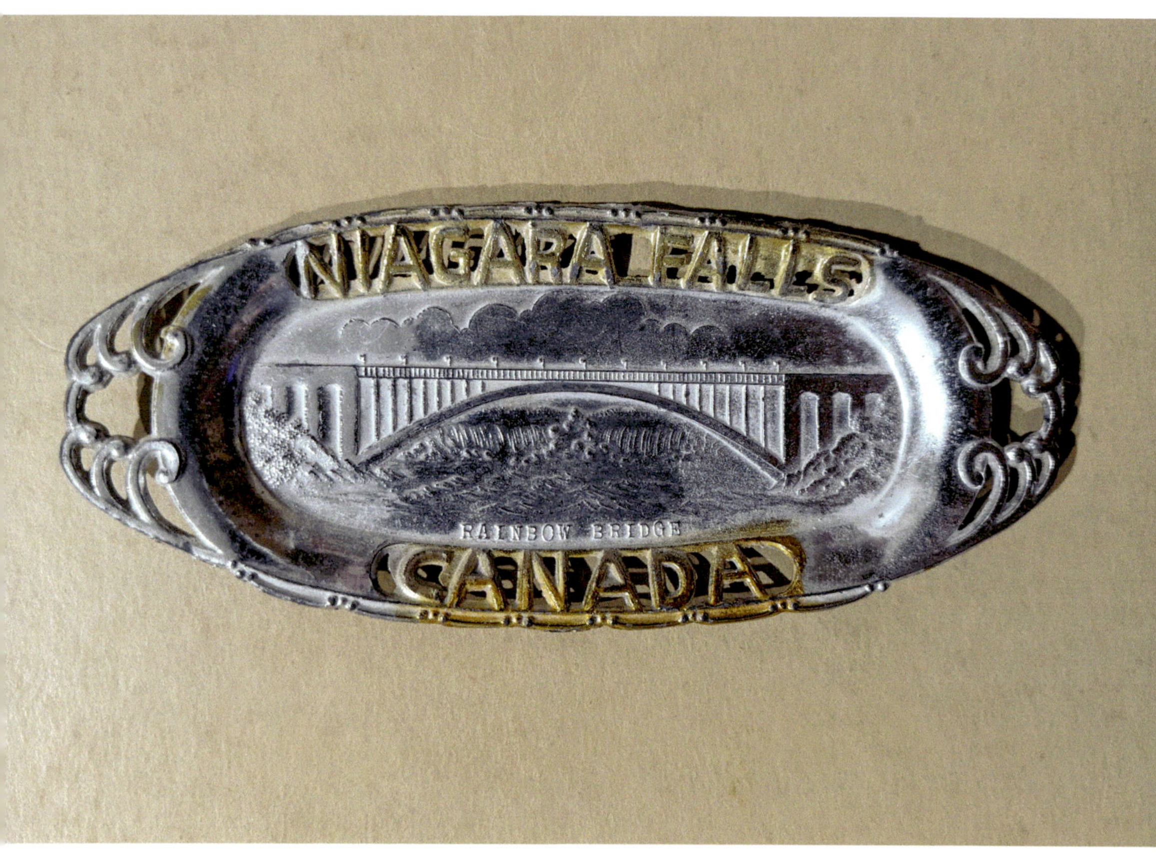

This book will document the amazing, beautiful bridges of the Niagara Falls area chronologically in photographs, engravings, sketches, and maps.

Below is a summary of some of the main bridges in the Niagara Falls area, from south to north.

Grand Island Bridges

In 1929, the Niagara Frontier Bridge Commission was formed by State of New York to study and implement a pair of bridges connecting Buffalo and Niagara Falls via crossings over the Niagara River at Grand Island. Such a crossing would save motorists time and gas. Previously, traveling between Buffalo and Niagara Falls had meant taking a roundabout road along the east side of the river.

The New York City-based engineering firm of Waddell & Hardesty was selected to perform a study on the proposed bridges, in collaboration with well-known architect Cass Gilbert (New York City's Woolworth Building) and Buffalo engineering firm George C. Diehl, Inc. Waddell & Hardesty had been founded by Dr. John Alexander Low Waddell in 1887. Born in Ontario, Canada, Waddell had founded the firm in Kansas City, Missouri, but moved to New York in about 1917 as the demand for new infrastructure began to heat up. The firm made its name on a variety of fixed and movable bridges, including the iconic Arroyo Seco Bridge in Pasadena, California, in 1913. J. A. L. Waddell partnered with his son, Needham Everett Waddell, for a few years until about 1920 and then formed a partnership in 1926 with his longtime associate, Shortridge Hardesty.

Once the study was presented to the commission, the firm designed the two bridges, which were completed in 1935. The North and South Grand Island Bridges are cantilever trusses that carry what is today Interstate 190 between Tonawanda, Grand Island, and Niagara Falls. The total cost of the two was less than $10 million. After a few decades, with growing traffic, there was a need to build twins for each of the bridges and make one for northbound and one for southbound vehicles. The new twin bridge to the existing South Grand Island Bridge opened in 1963 and the new twin bridge to the North Grand Island Bridge opened in 1965, both designed by Hardesty & Hanover.

Rainbow Bridge and the Niagara Falls Bridges

The first bridge on the site closest to Niagara Falls proper was built in 1868 but was destroyed by high winds in 1889. It was rebuilt a few months later. It was taken down in 1898 and replaced with the Falls View Bridge, a steel arch bridge that, living up to its name, offered an excellent view of the falls. The toll for this bridge as of 1937 was 5 cents for pedestrians, and free for children under five; 25 cents for a car and driver, and 5 cents for each additional occupant. The toll for a truck was either 35 or 50 cents, depending on its weight.

On January 27, 1938, the old steel arch Falls View Bridge crashed into the river under the tonnage of its own ice accumulation. It needed to be replaced quickly. That year, a joint resolution of Congress created the Niagara Falls Bridge Commission, which has representation from both the United States and Canada. The commission was created specifically to fund, construct, and operate the new bridge (but continued to be a useful entity and is still in existence today). The old bridge had carried 1,148,208 vehicles in 1937, with over 95 percent of those being passenger cars, and the new bridge was projected to carry more than 1.2 million vehicles a year.

In the 1930s, Waddell & Hardesty was contracted to design the new bridge (Aymar Embury II was the bridge architect; bridges are often designed only by engineers, but in some cases for more high visibility crossings, architects are brought on to work out some of the aesthetic details which engineers can incorporate into the structural design). Due to the height of the banks above the water, a movable bridge was not necessary; there would be plenty of clearance for any boats to pass underneath. Given the nature of the site, the structure type selected was an arch bridge, which had worked well for the Whirlpool Rapids Bridge further north. Both structurally and aesthetically, it would be a great option.

Firm founder J. A. L. Waddell died in 1938, so the firm was led by his partner, Shortridge Hardesty, who personally spearheaded the design efforts on the new international bridge, which was to have a main span that was a hingeless arch 960 feet in length. In 1939, King George VI and Queen Elizabeth visited Niagara Falls to dedicate the site of the new bridge, which was about 550 feet north of the previous bridge. Construction on this Rainbow Bridge was begun on May 4, 1940, and the ribbon cutting was held on November 1, 1941. A host of dignitaries was present for the occasion. The bridge was generally well received by critics. It was awarded a First Place Class A prize from the American Institute of Structural Steel. The Rainbow Bridge instantly became an ubiquitous symbol of the falls, appearing in countless postcards and souvenirs taken home by millions of visitors from around the globe, not to mention countless snapshots. The Rainbow Bridge was a high point of Shortridge Hardesty's engineering career; when he died in 1956, it was the first thing mentioned in his obituary.

Whirlpool Rapids Bridges

In 1846, two companies were formed to build a bridge over the Niagara River near the falls—the Niagara Falls Suspension Bridge Company (Canadian) and the International Bridge Company (American). The next year, the companies commissioned Charles Ellett Jr. to build a bridge at the narrowest point of the Niagara gorge. The challenges to accomplishing this were numerous, starting with the question as to how to begin. Cable would have to get across the gorge, but how?

A kite-flying contest was selected as the best way to do this, and though on the first day no one succeeded in getting a kite across the gorge, on the second day a boy named Homan Walsh managed to fly his kite to the opposite side. A heavier rope was attached and pulled across, followed by a 1,190-foot-long wire cable that was anchored to a 50-foot-high wooden tower on each side. Rides across this iron cable in an iron rocking chair-like basket were used to transport equipment and laborers from one side to the other, but were also offered to the general public for the princely sum of a dollar. Over 100 people a day took the ride. Construction was harrowing and dangerous but the Niagara Falls Suspension Bridge, the very first bridge across the river near the falls, opened to the public on August 1, 1848.

Next, future Brooklyn Bridge engineer John A. Roebling designed the Niagara Falls Railway Suspension Bridge at this site, which opened in 1855 and carried both rail and foot traffic. The bridge was updated and redesigned by Leiffert L. Buck and completed in 1886, but within ten years this bridge too became outdated and its replacement, the Whirlpool Rapids Bridge, was built in 1896–97. Also designed by Buck, it was a two-hinged spandrel-braced arch-type structure. The existing bridge remained in service while the new one was constructed. The Whirlpool Rapids Bridge still stands today.

Just upstream from the Whirlpool Rapids Bridge is the abandoned Michigan Central Railway Bridge (no longer in use), an arch structure built in 1924–25. The single-track bridge is owned by Canadian Pacific Kansas City (CP Rail) which purchased it in 1990. It replaced the Niagara Cantilever Bridge that crossed there from 1883 to 1925.

Lewiston-Queenston Bridges

A suspension bridge between Lewiston, NY, and Queenston, Ontario, was first built in 1851, but this Queenston-Lewiston Bridge was destroyed by wind not long after. Its replacement was another suspension bridge, which was moved to the site from the vicinity of the Rainbow Bridge in 1898 when the Upper Steel Arch Bridge was completed. By the mid-twentieth century, it was time to build a newer bridge at that point along the river. The firm of Waddell & Hardesty—now known as Hardesty & Hanover (in 1945, Shortridge Hardesty partnered with Clinton Hanover to form the firm, still in existence today as H&H)—was contracted *c.* 1960 to design the new bridge. The new

Lewiston-Queenston Bridge was similar in style to the Rainbow Bridge. It cost $16 million to build. The fabricator was the Bethlehem Steel Company. The new bridge was under construction from 1960–62 and opened in late 1962. It won the Long Span Prize Bridge Award (for bridges with spans of 400 feet or more) from the American Institute of Structural Steel in 1962.

As of 2021, the total assets of the Niagara Falls Bridge Commission were $473 million. Their bridges (the Rainbow, Whirlpool Rapids, and Lewiston-Queenston bridges) are inspected regularly under a detailed inspection program where critical areas of the bridges are "hands-on" inspected to ensure the good condition of critical elements of the bridges. Photos are taken to document any conditions noted, and reports written describing the results of the in-depth inspection.

According to the NFBC website: "No other single factor contributes more to tourism growth—indeed the export trade economy overall—than the NFBC's three monumental linkages of concrete and steel. Niagara Falls crossings rank second only to Kennedy International as the busiest port of entry between the U.S. and Canada. If you are about to travel over Niagara's gorge, your vehicle will take one of about 7.2 million passages expected this year on NFBC bridges."

I hope you will enjoy this scenic ride through the many bridges of Niagara Falls and look with wonder at the beautiful juxtaposition of these engineering marvels with one of nature's wonders.

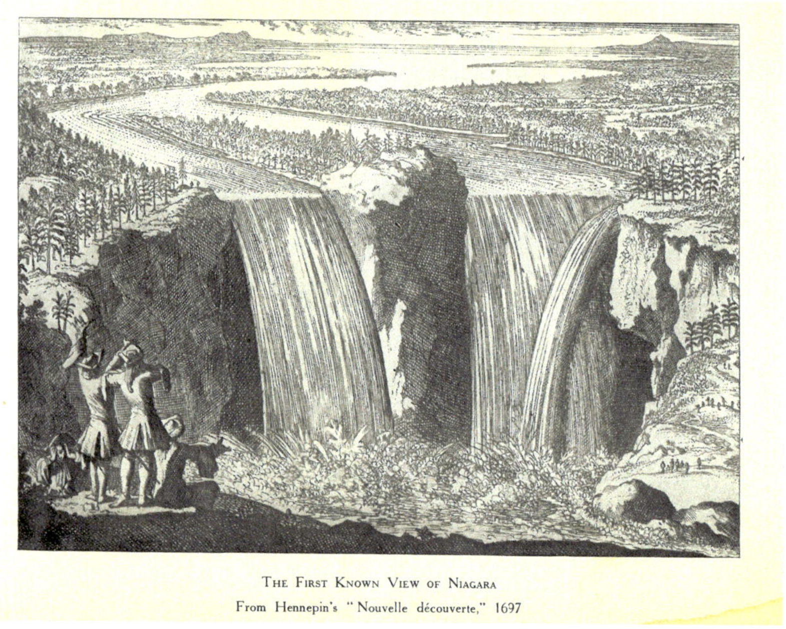

The first known view of Niagara Falls, published in Frenchman Louis Hennepin's (1640–1705) book, *Nouvelle Decouverte* (*New Discoveries*).

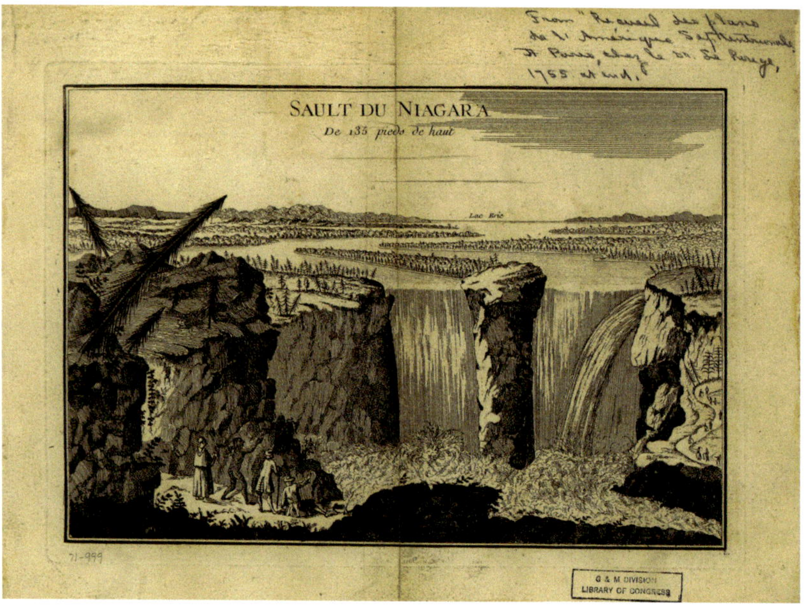

This engraving of Niagara Falls was made by Georges-Louis le Rouge (born 1712), royal geographer to King Louis XV, and published in *Recueil des Plans de l'Amérique Septentrionale* (*Collection of the Maps of North America*) in 1755. The engraving depicts a view of the falls from an elevation of 135 feet. In the foreground are four men: three Europeans, one of whom is in clerical garb holding a cross, and a Native American, who is possibly their guide. (*Library of Congress*)

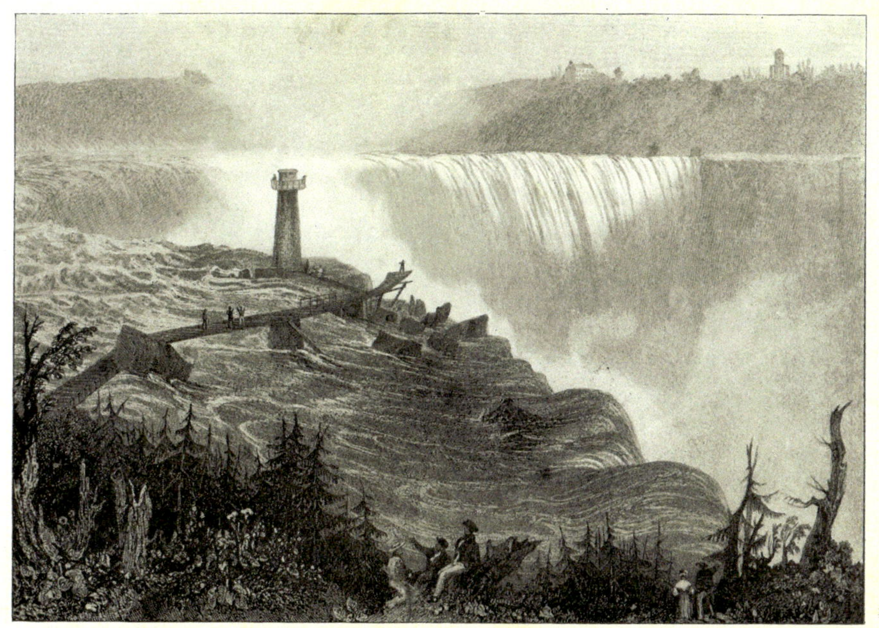

THE HORSE SHOE FALL, NIAGARA — WITH THE TOWER
Painted by W. H. Bartlett; engraved by P. Brandard. Published in London by George Virtue, 1837

An 1837 view of the Horseshoe Falls, painted by W. H. Bartlett and engraved by P. Brandard. Several onlookers can be seen crossing a small bridge leading to the edge of the falls.

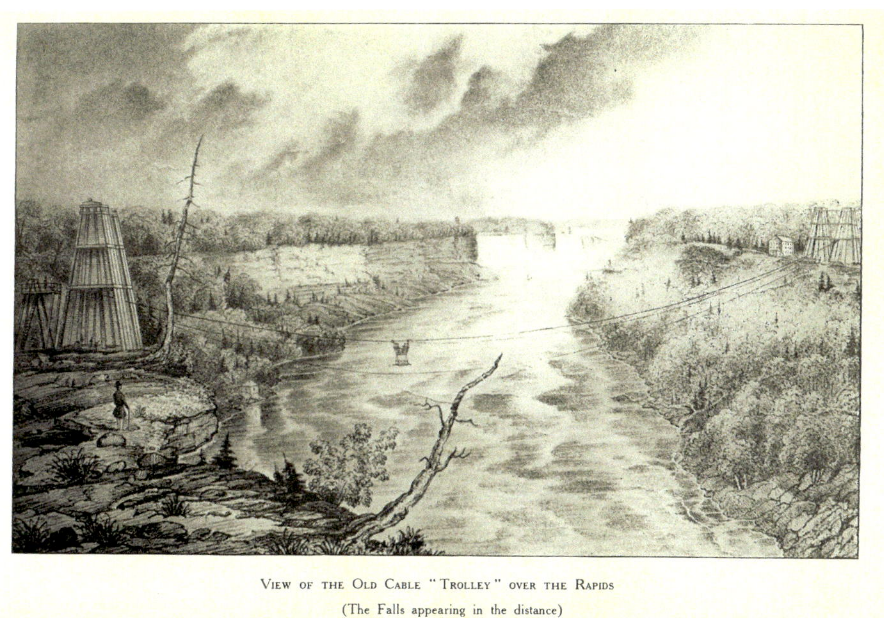

VIEW OF THE OLD CABLE "TROLLEY" OVER THE RAPIDS
(The Falls appearing in the distance)

A depiction of the suspension cable trolley that was in place while the Niagara Falls Suspension Bridge was under construction, *c.* 1848.

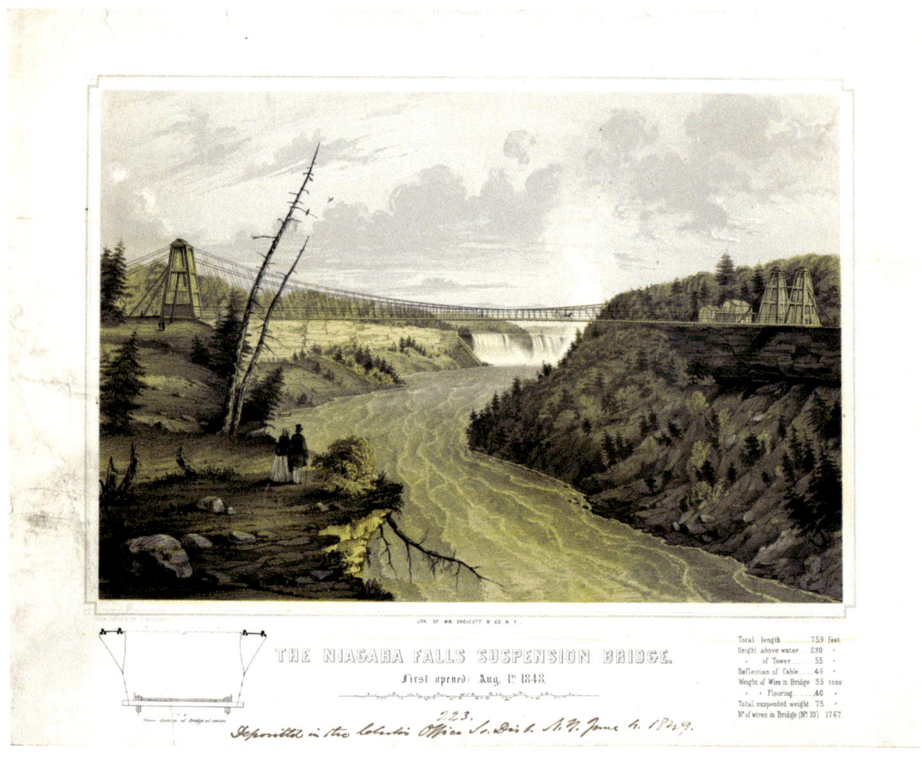

A 759-foot-long suspension bridge over the Niagara River opened on August 1, 1848. It was 230 feet above the water, with a tower height of 55 feet. (*Library of Congress*)

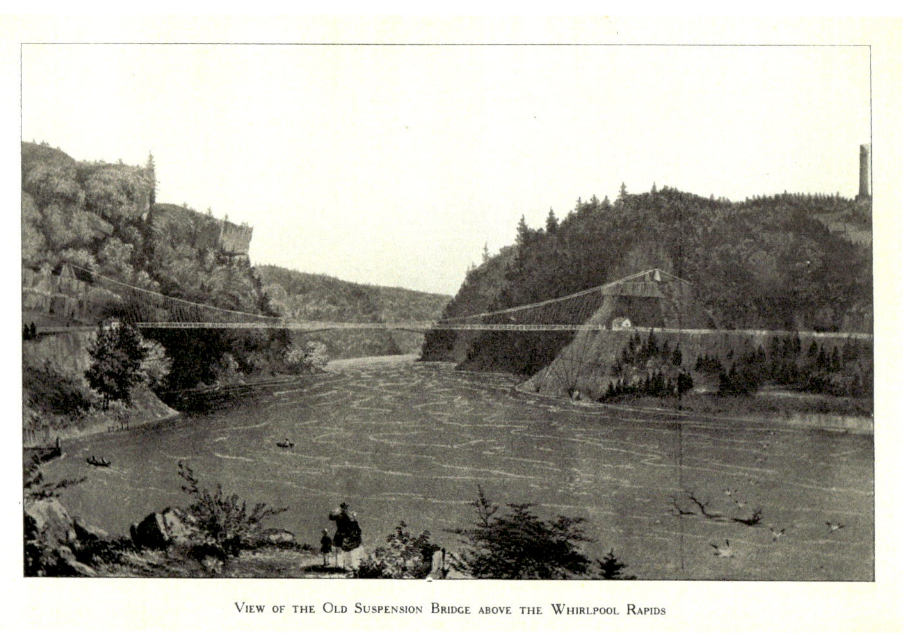

An 1850s view of the suspension bridge over the Niagara River.

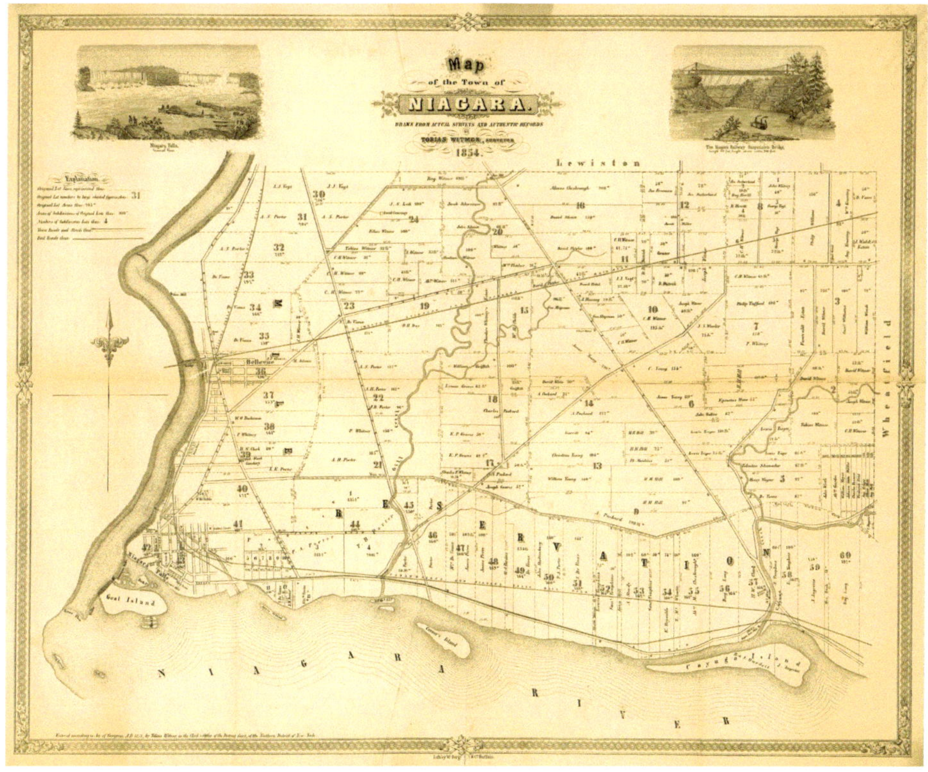

An 1854 township cadastral map of Niagara Falls, New York, showing original survey lots/lines/ numbers, contemporary property tracts/lines/numbers, tract acreages, and landowners' names. The railroad suspension bridge is visible, as are the bridges to Goat Island from the mainland. (*Library of Congress*)

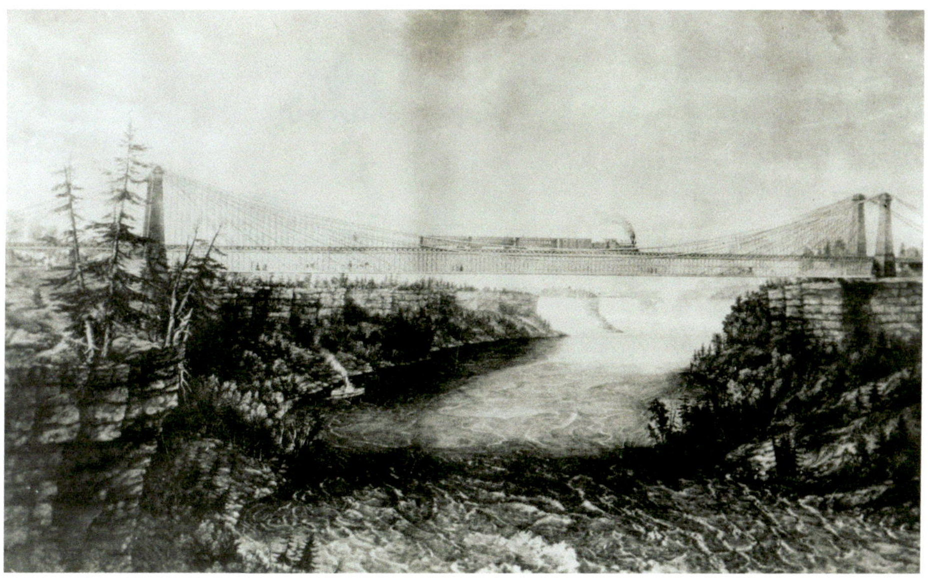

An 1855 image of the suspension bridge features a steam train crossing it. (*National Archives*)

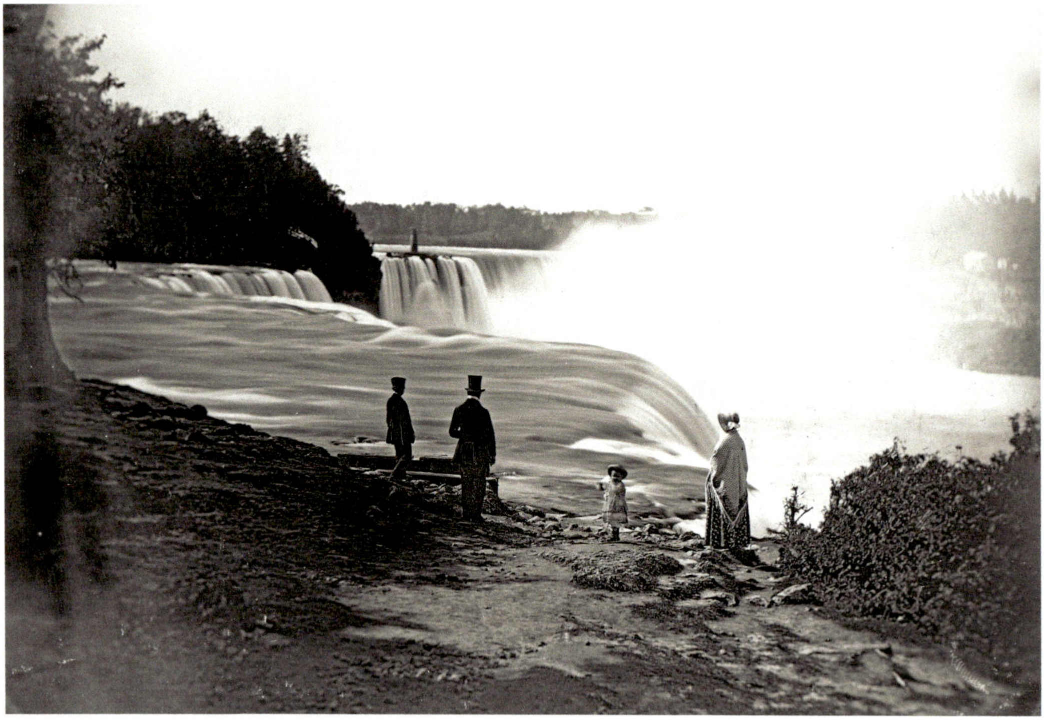

A *c.* 1854–55 daguerreotype of the America Falls seen from Prospect Point. This is probably among the oldest known photographs of Niagara Falls. (*US Geologic Survey*)

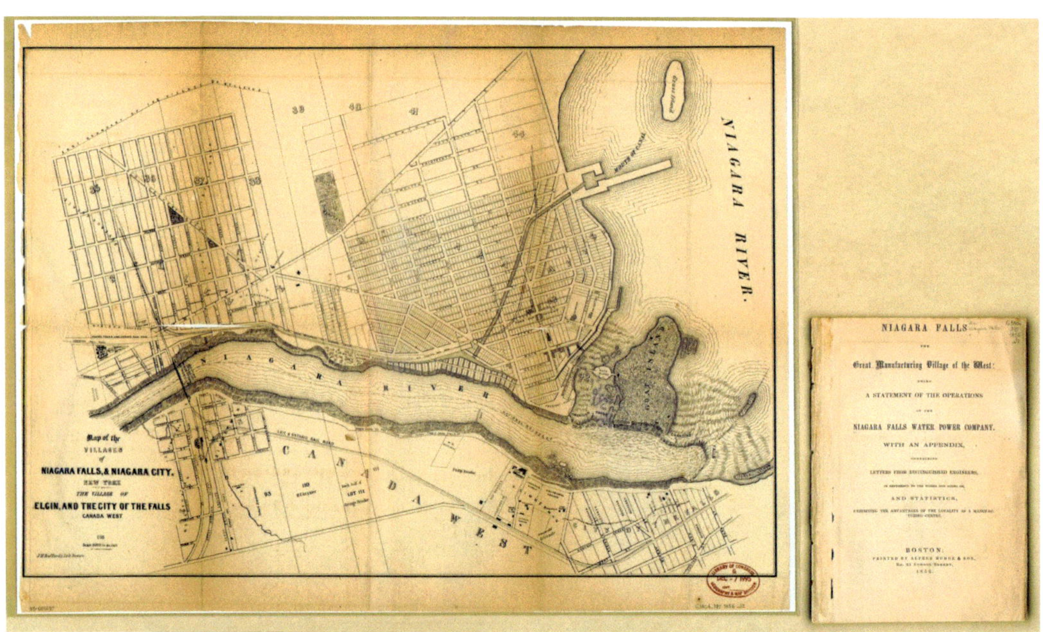

An 1856 map of the villages of Niagara Falls and Niagara City shows the Railway Suspension Bridge that was built a few years prior, as well as the Goat Island bridge. (*Library of Congress*)

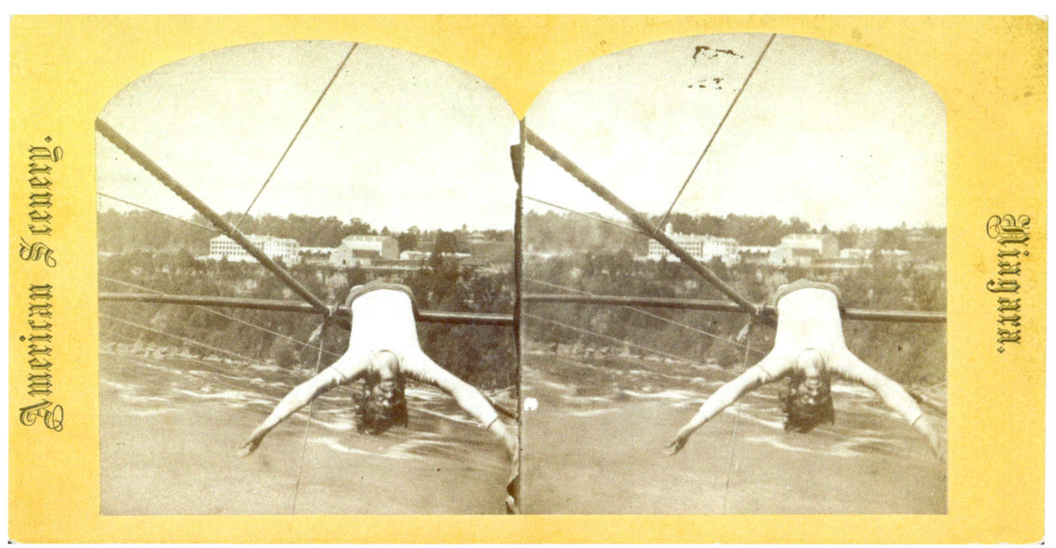

Bridges were not the only way to cross the Niagara River. Some daredevils attempted to cross on nothing more than a tightrope. In 1859, the acrobat Charles Blondin (aka Jean Francois Gravelet), became the first person to walk across the falls on a tightrope. Over the next two years, he repeated this feat several times.

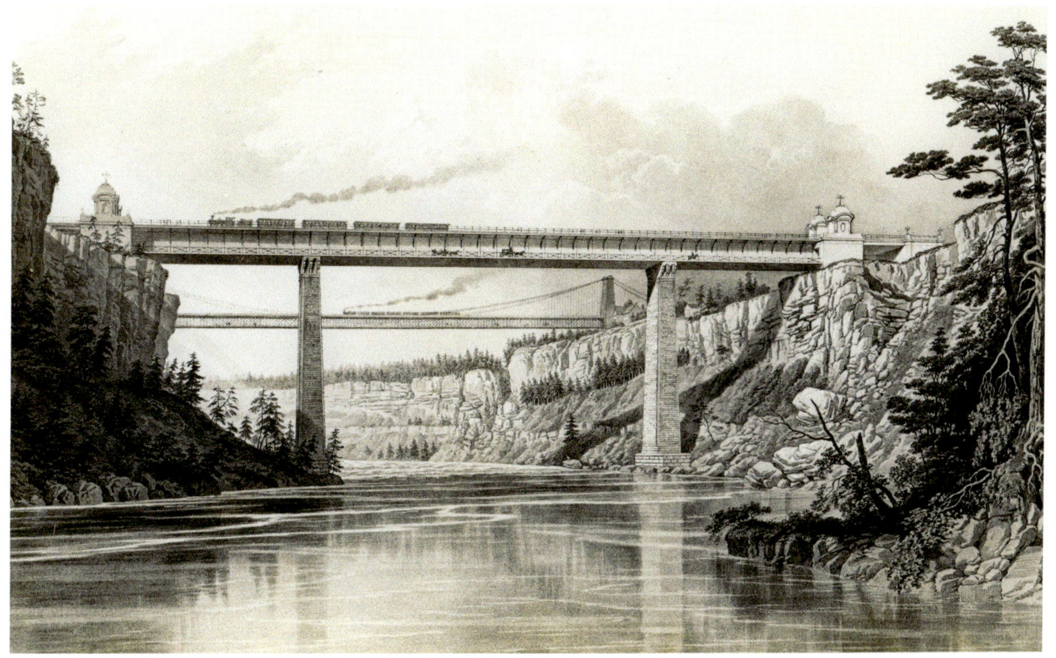

A proposed "tubular bridge" to cross the Niagara River, as shown in an 1860 publication called *Construction of the Great Victoria Bridge in Canada*. (*National Archives*)

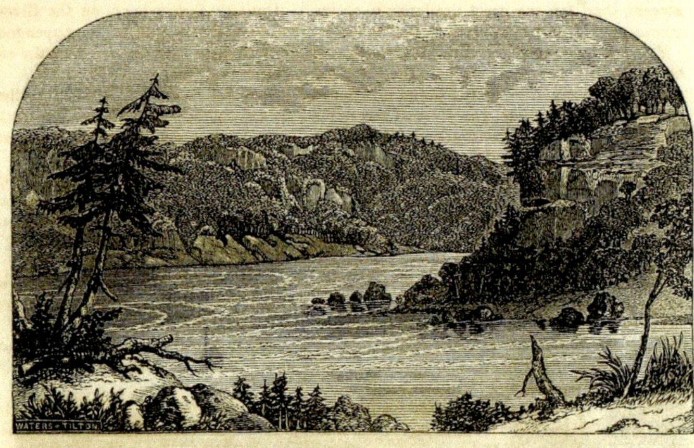

THE WHIRLPOOL—NIAGARA.

As stated on a previous page, the best view of the Whirlpool is to be had from the edge of the river, on the American side, and to give an idea of this scene we present an accurate representation of it, taken from the point of the angle, where the river, after proceeding to the point indicated in the far-off corner of the above sketch, whirls round, and finds its outlet down the river in the foreground, on its way to Lake Ontario. (See remarks on a previous page.)

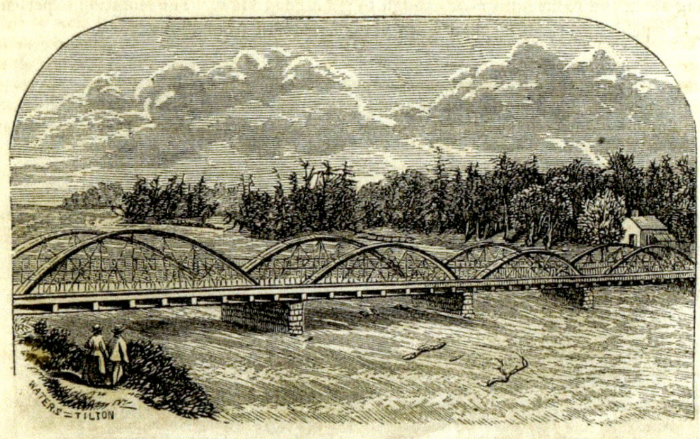

THE BRIDGE LEADING TO BATH AND GOAT ISLANDS.

The above sketch represents the well-known bridge which spans the river to Bath Island, and from thence leads across another small bridge to Goat Island. At Bath Island passengers pay the toll of 25 cents, which admit them to cross and recross during all the season. The bridge is not a suspension one, although similar to such in appearance. It is built on three piers, founded in the bed of the river by means of cribs filled with heavy masonry, and is altogether a graceful and substantial erection, strong enough for all the traffic passing across it, and for resisting the powerful current of the rapids as they rush down and flow under it on their way over the American Fall.

15

A mid-nineteenth-century view of the bridge leading to Bath Island (now known as Green Island). At Bath Island, pedestrians paid a 25-cent toll, which allowed them to cross and recross for the entire season. The bridge was built on three piers on foundations in the river by means of cribs filled with heavy masonry.

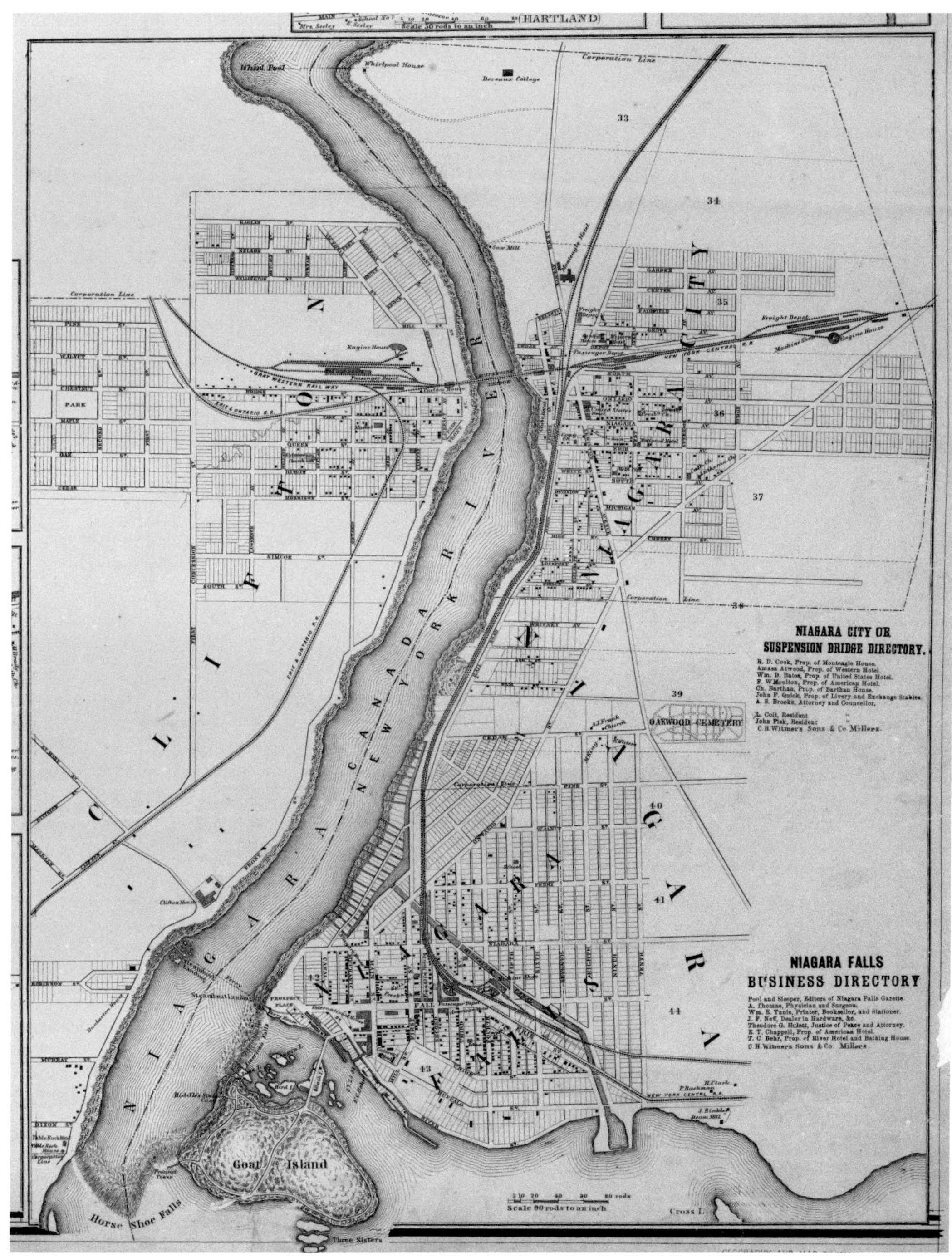

An 1860 township cadastral map of both sides of Niagara Falls, showing original survey lots/lines/ numbers, contemporary property tracts/lines/numbers, tract acreages, and landowners' names. The railroad suspension bridge and Goat Island bridges are visible. The American side is more developed than the Canadian side at this point. (*Library of Congress*)

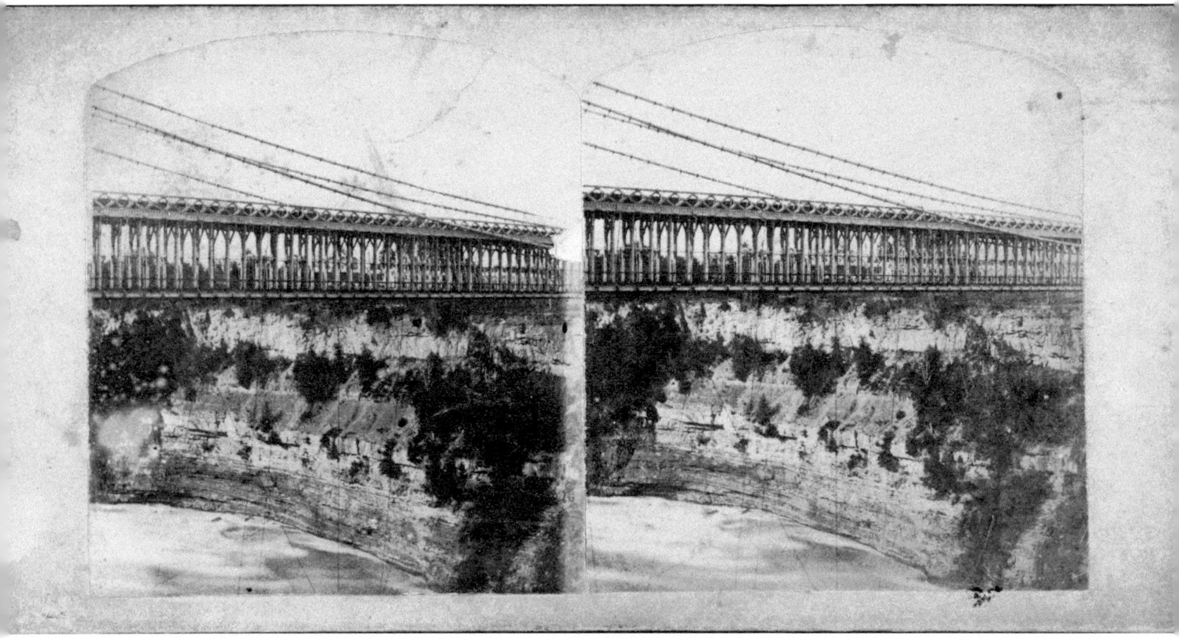

At top, the suspension bridge in an *c.* 1860 stereoscopic view; at bottom, the reverse of the card gives details of the bridge's construction. The stereoscope was a popular mid- to late-nineteenth-century optical viewing device that allowed people to experience scenic images as three-dimensional. (*Library of Congress*)

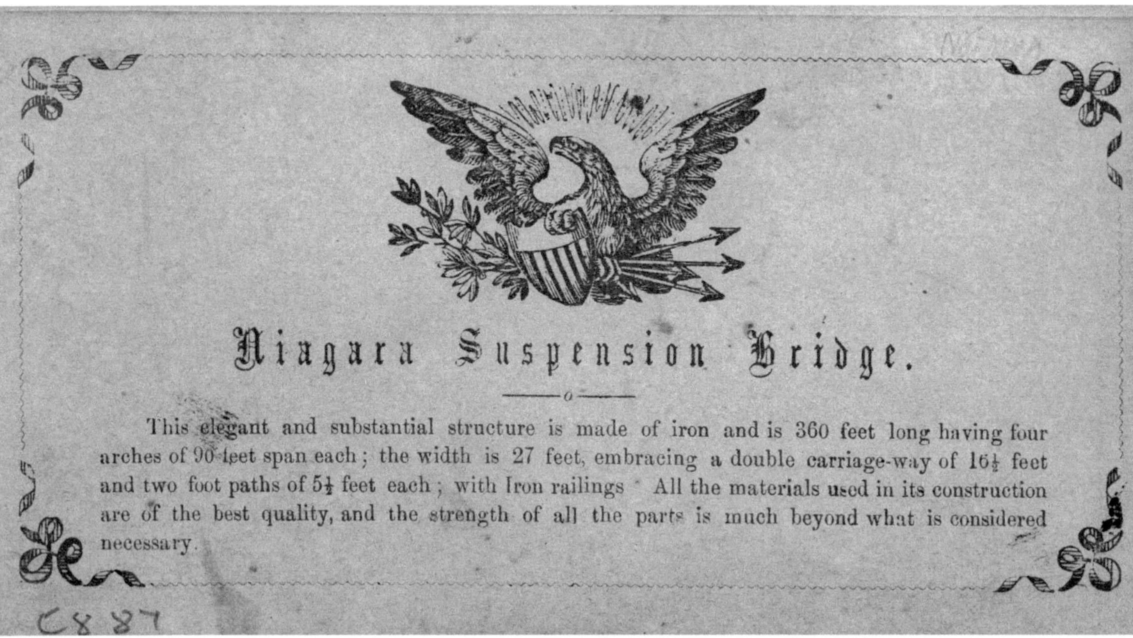

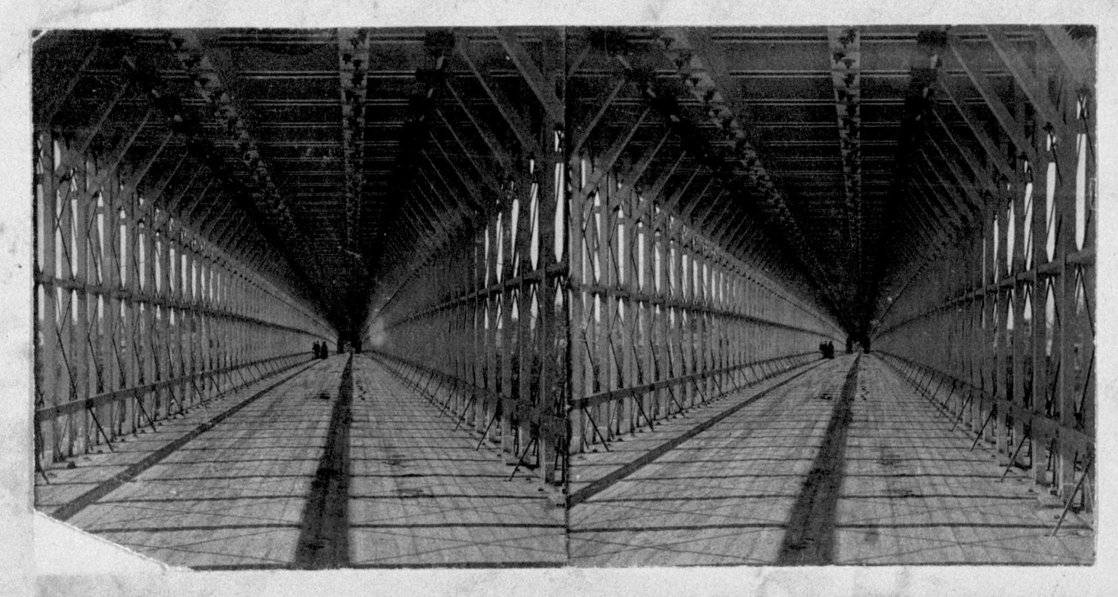

The interior of the pedestrian level of the suspension bridge as seen on a *c.* 1860 stereoscopic view. The reverse touts the unique view from the interior of the bridge. (*Library of Congress*)

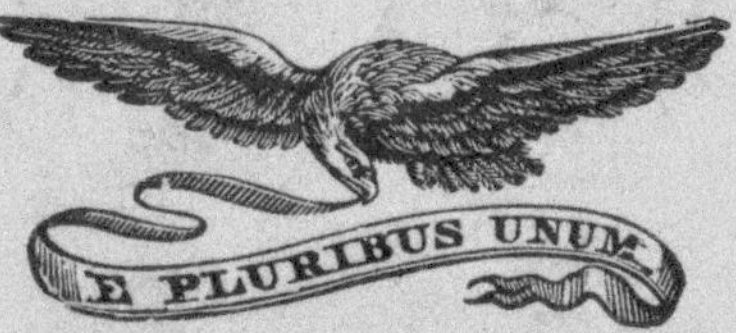

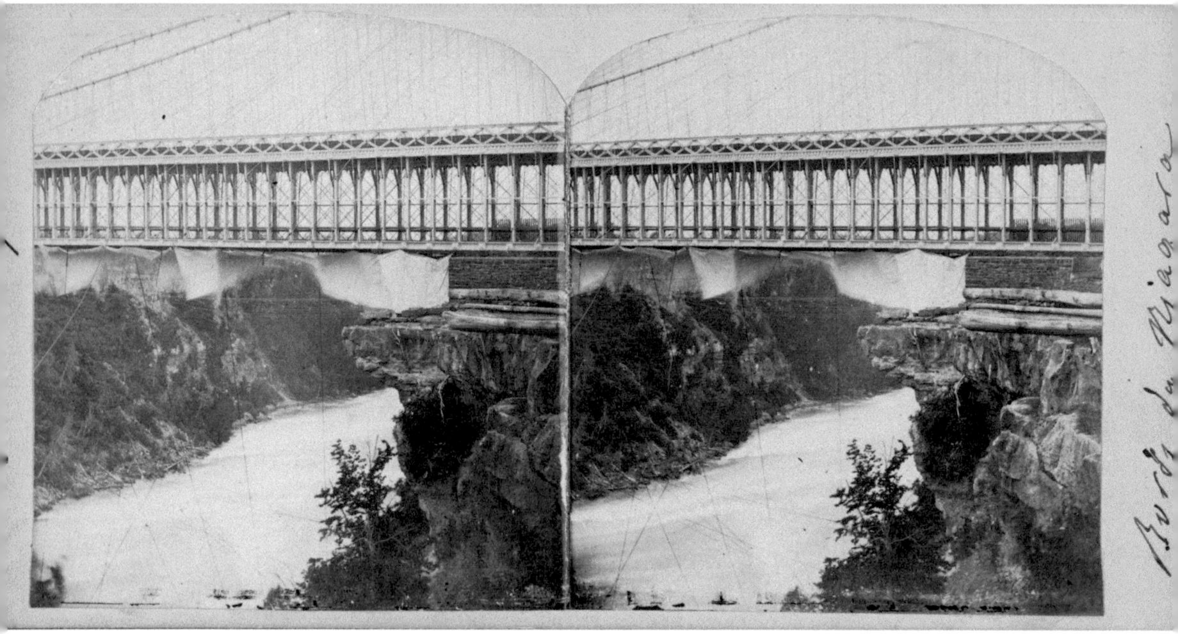

A *c.* 1860–65 stereoscope view of the suspension bridge at Niagara Falls. (*Library of Congress*)

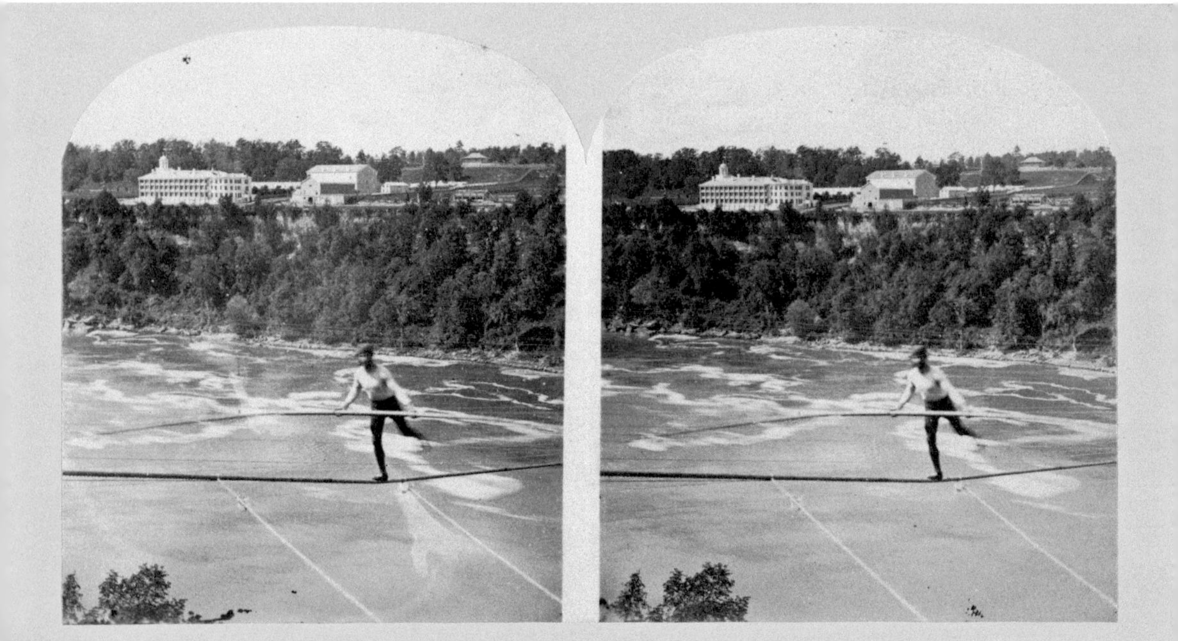

Tightrope walker crossing over the Niagara River, *c.* 1860s. (*Library of Congress*)

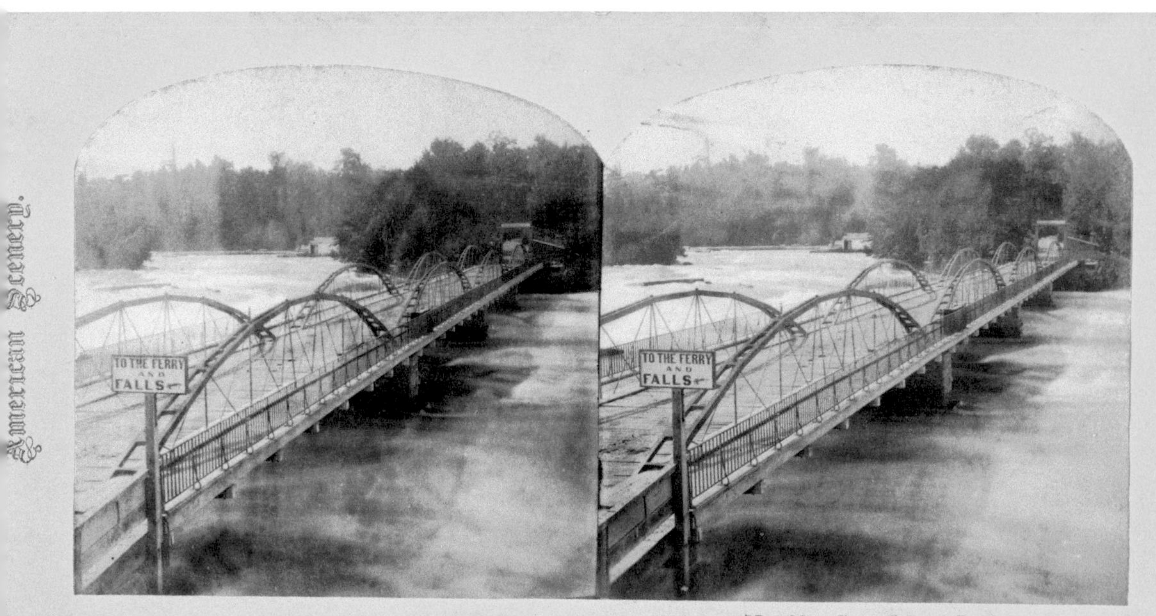

A *c.* 1865 photo of the Goat Island Bridge. The islands at Niagara Falls offer amazing views of the turbulent waters, so constructing bridges from the mainland and between islands was a priority in the nineteenth century. (*Library of Congress*)

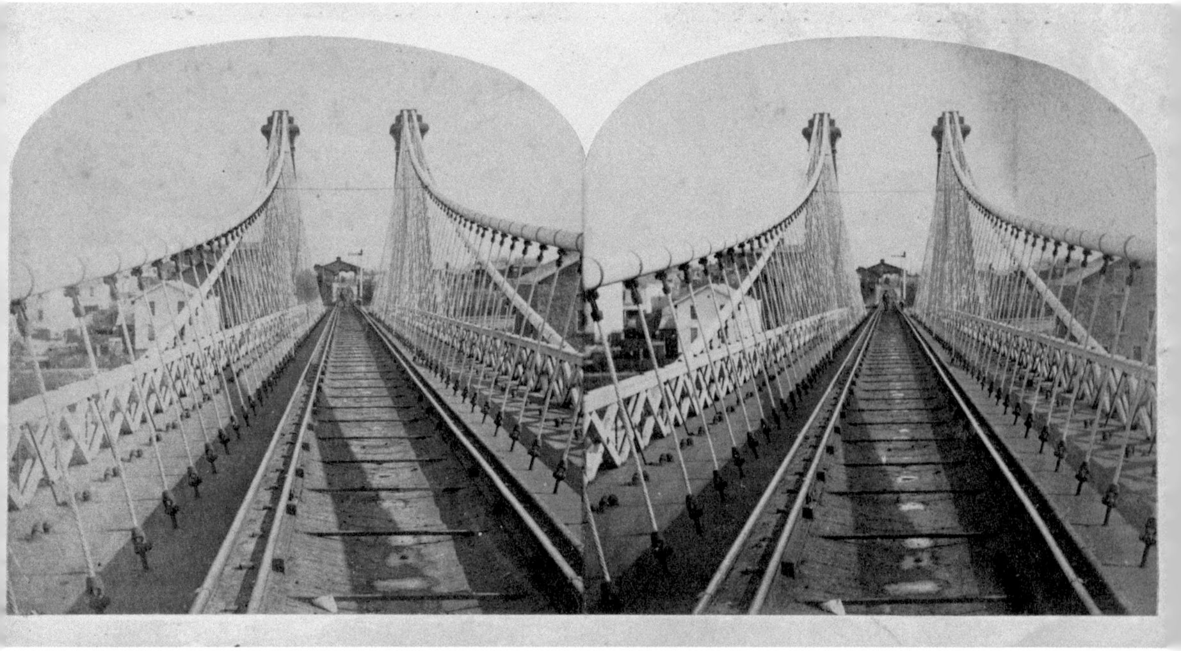

The Niagara Falls Suspension Bridge. Top view from the middle, looking toward the American side *c.* 1860s.

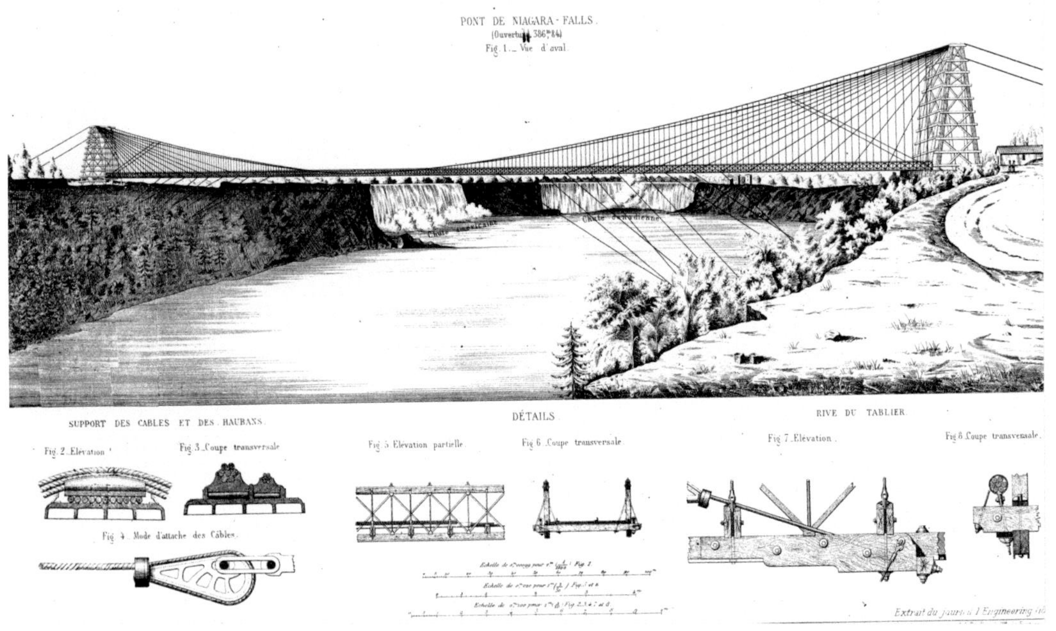

A page from an 1870 French engineering book shows details of the Niagara Falls suspension bridge. (*National Archives*)

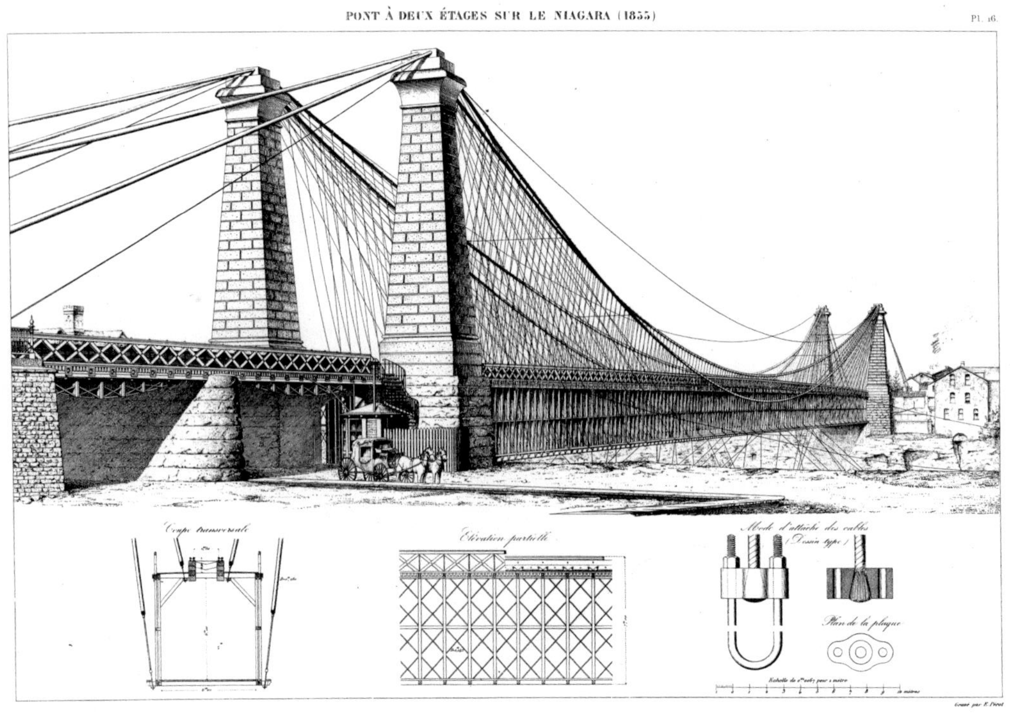

The Niagara Falls Suspension Bridge, built 1855, as seen in an 1870 publication. (*National Archives*)

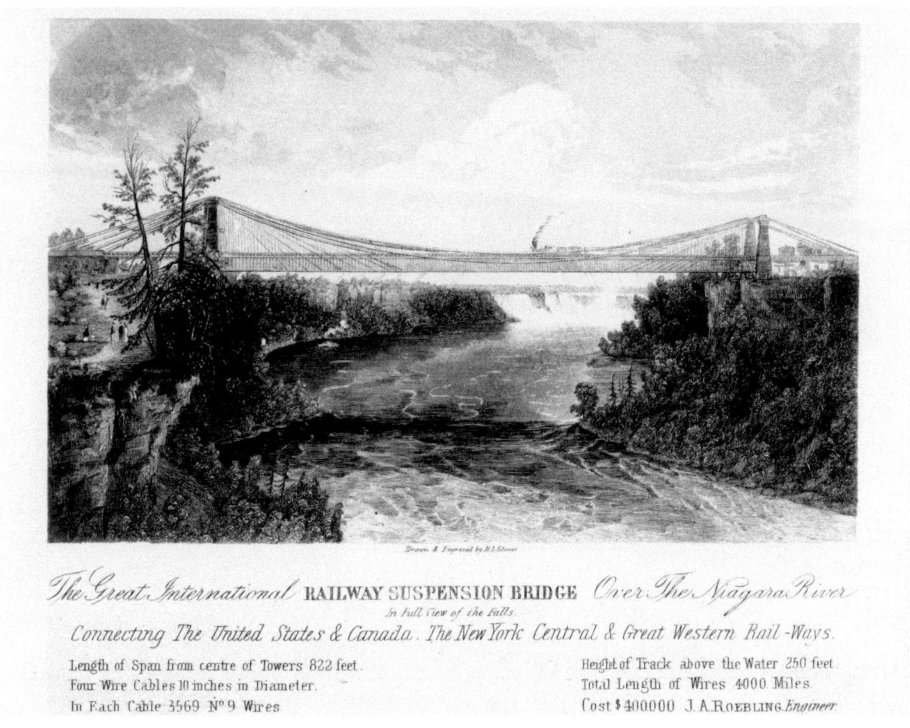

An 1871 engraving of the Railway Suspension Bridge. The span between towers was 822 feet. (*National Archives*)

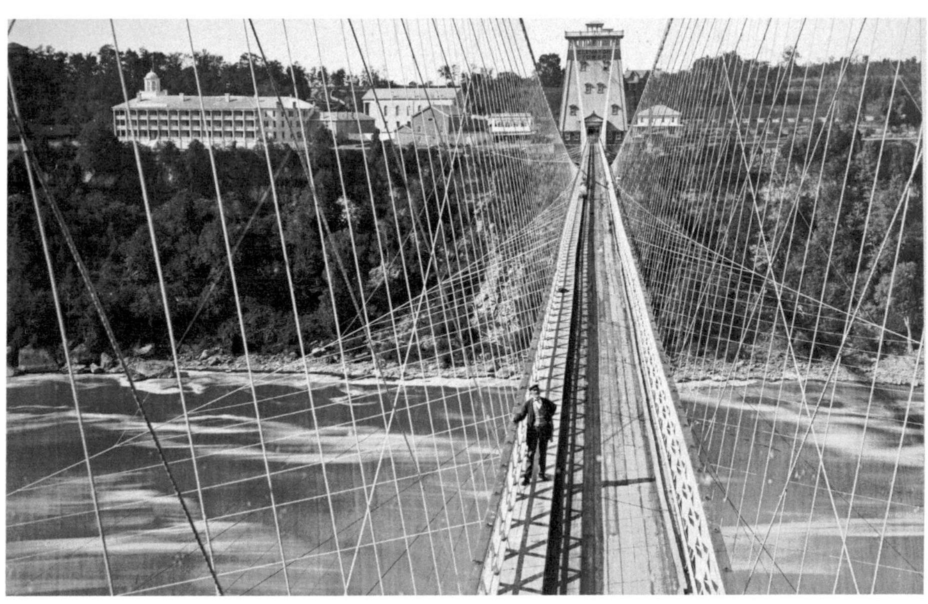

This photograph shows a man standing on the Falls View suspension bridge (*c*. 1875) with the elevator tower on the Canadian side visible in the distance. The bridge was very narrow at only 10 feet in width. The bridge was completely destroyed by a storm in January 1889 and was replaced with a wider duplicate bridge in May 1889. (*Library of Congress*)

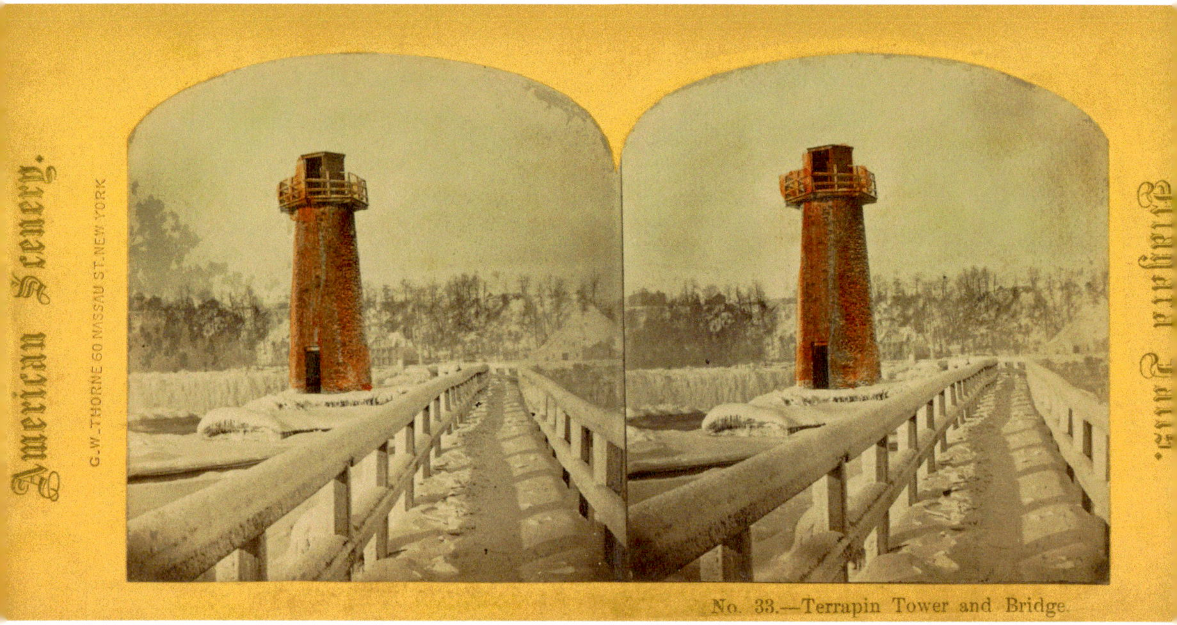

A stereoscopic view and a print both showing the Terrapin Tower (torn down in 1872) and bridge as seen from Goat Island, *c.* 1860s. Built *c.* 1830, the Terrapin Tower was strategically located to offer excellent views and was about 40 feet high. (*Library of Congress*)

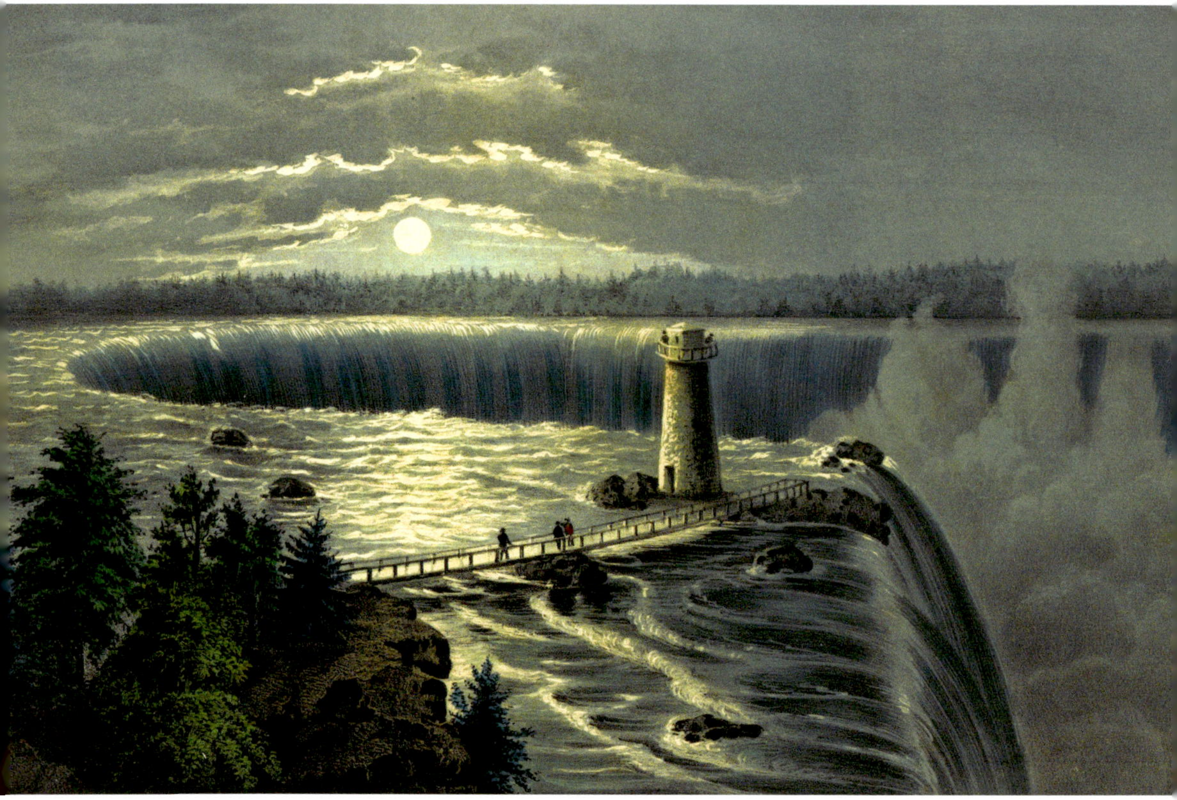

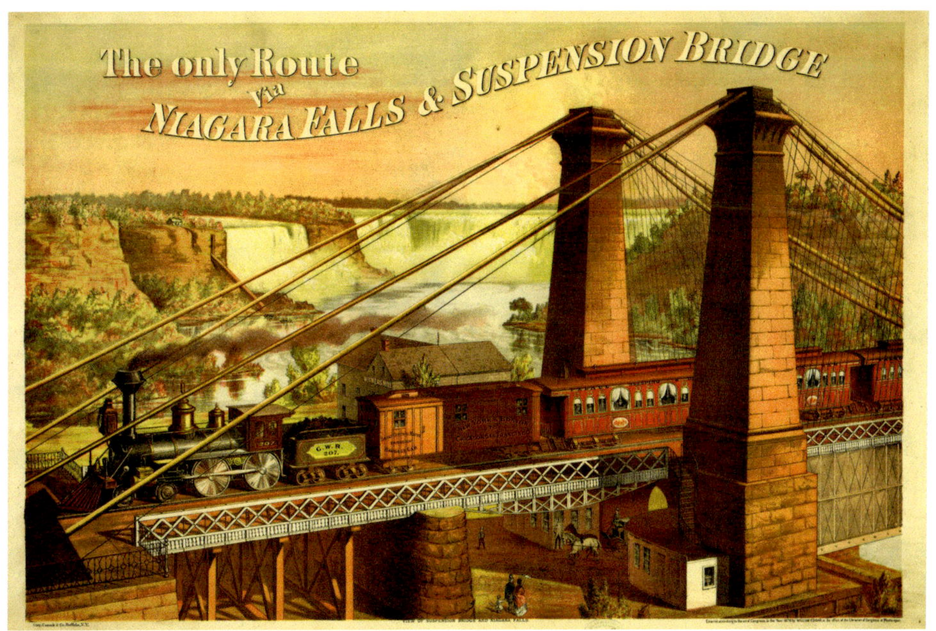

A *c.* 1876 lithograph of the Niagara Falls Suspension Bridge. Opened in 1855, the bridge was designed by the engineer John A. Roebling, who would go on to design the Brooklyn Bridge. (*Library of Congress*)

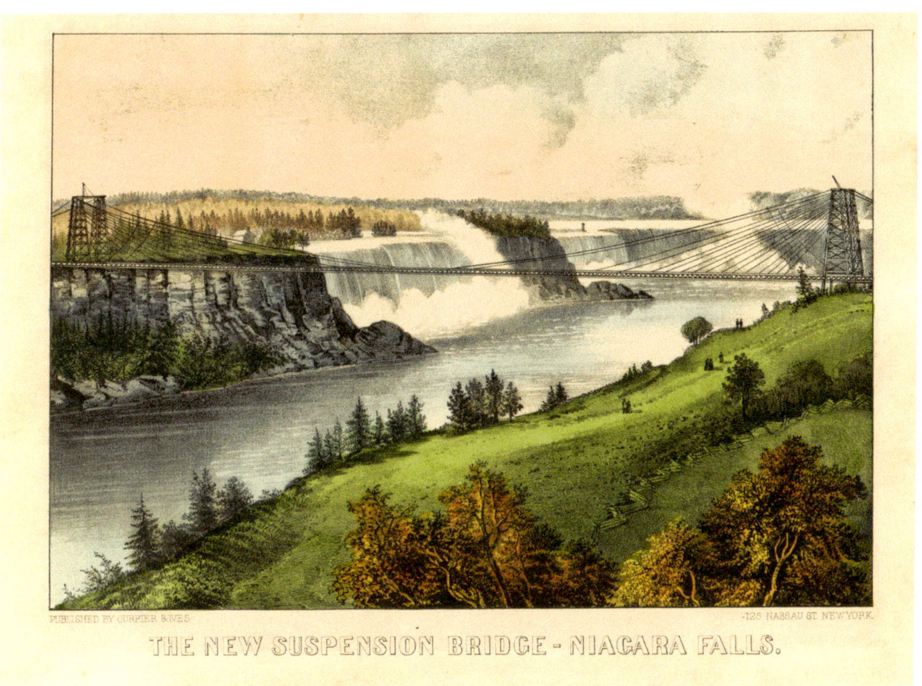

A nineteenth-century Currier and Ives view of the Niagara Falls suspension bridge. Currier and Ives was a famous New York printmaker that produced scenic, colorful lithographic views of America that were often framed and hung in homes across the country. (*Library of Congress*)

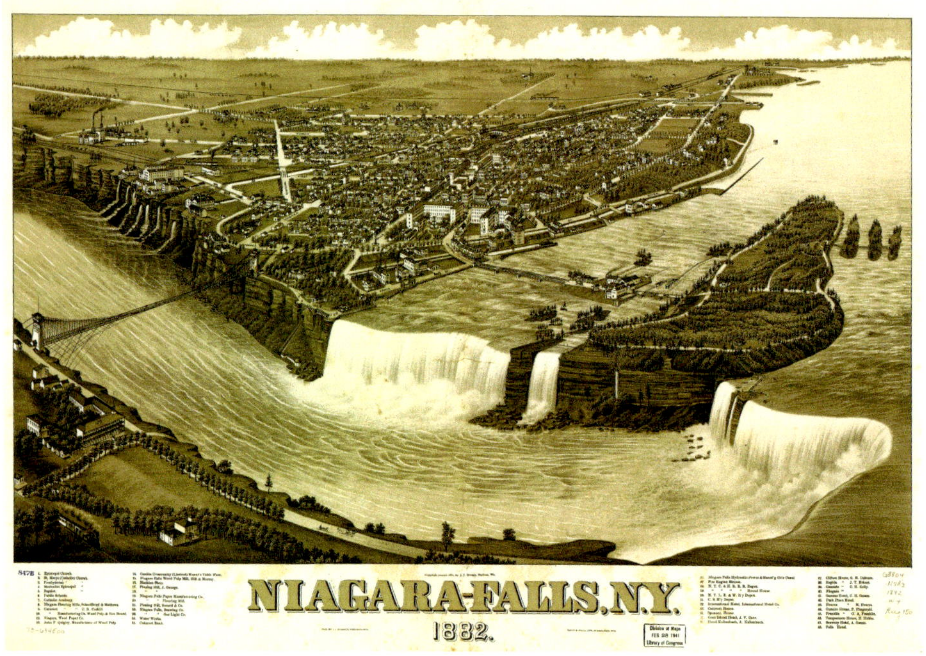

A bird's-eye view of Niagara Falls, New York, in 1882 (published by J. J. Stoner of Madison, Wisconsin) showing various points of interest. Before the days of aerial photography, bird's-eye views were the best way to get a scenic perspective of a tourist destination. (*Library of Congress*)

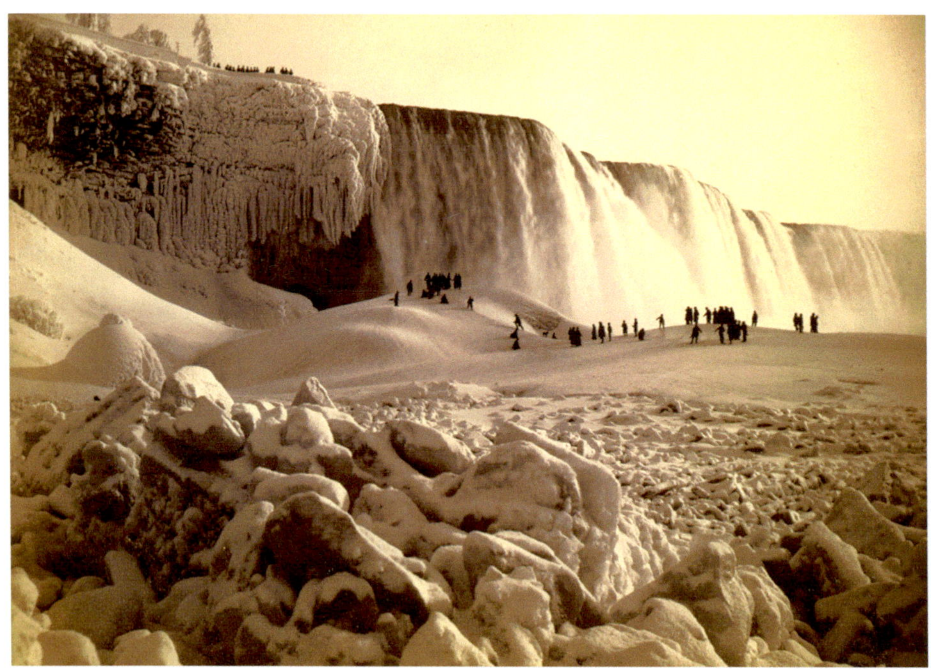

A great photograph (by George Barker) of the ice bridge and ice mound, with the American Falls in the background, 1883. (*Library of Congress*)

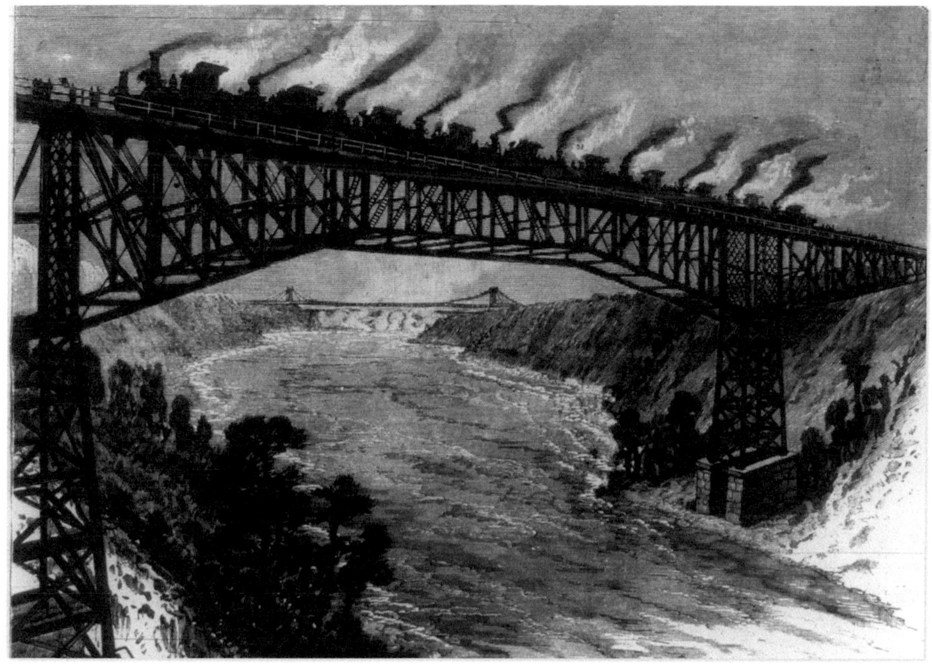

The opening of the Cantilever Bridge across the Niagara River, December 20, 1883 as seen in *Frank Leslie's Illustrated Newspaper*. This image shows the test trains, consisting of twenty-two locomotives and twenty-two loaded gravel cars crossing the bridge. (*Library of Congress*)

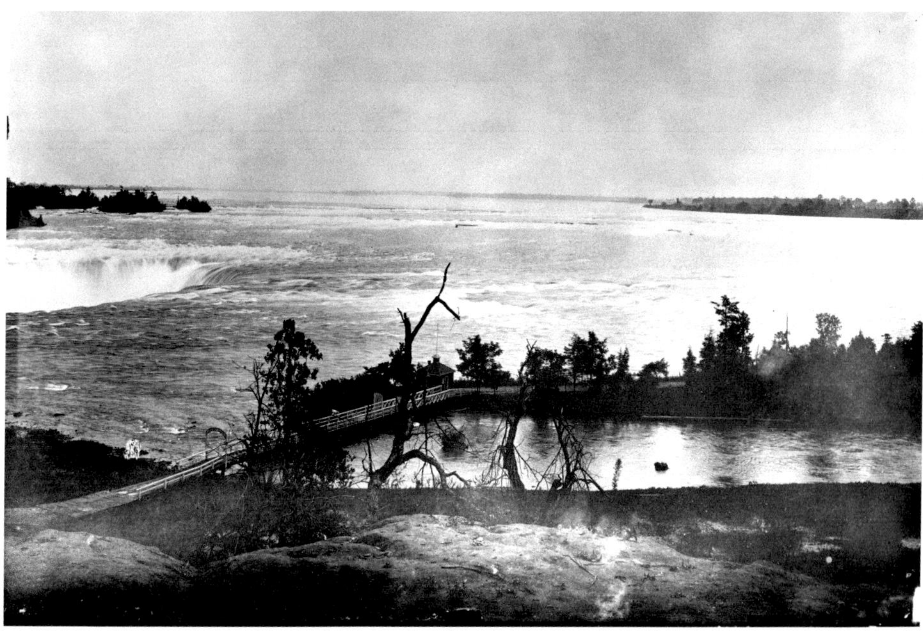

A view of the falls from the Canadian side, in an 1886 United States Geologic Survey photo. The USGS's main interest was documenting the unique geologic circumstances of the falls area, but in the process, they also recorded the bridges. (*US Geologic Survey*)

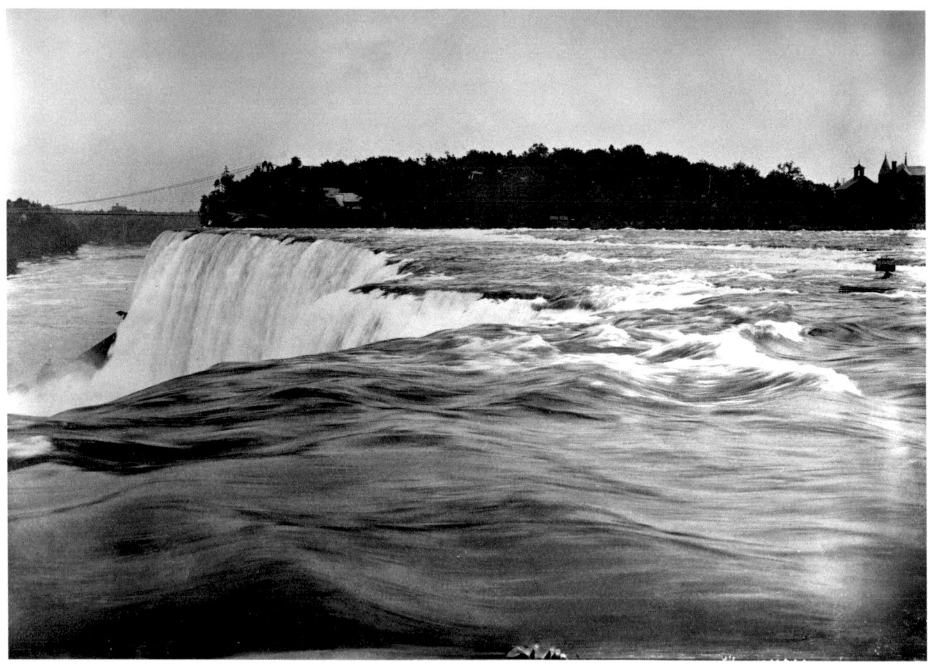

Rapids above the American Falls, as seen in 1886 from Goat Island. The suspension bridge is in the background at left. (*US Geologic Survey*)

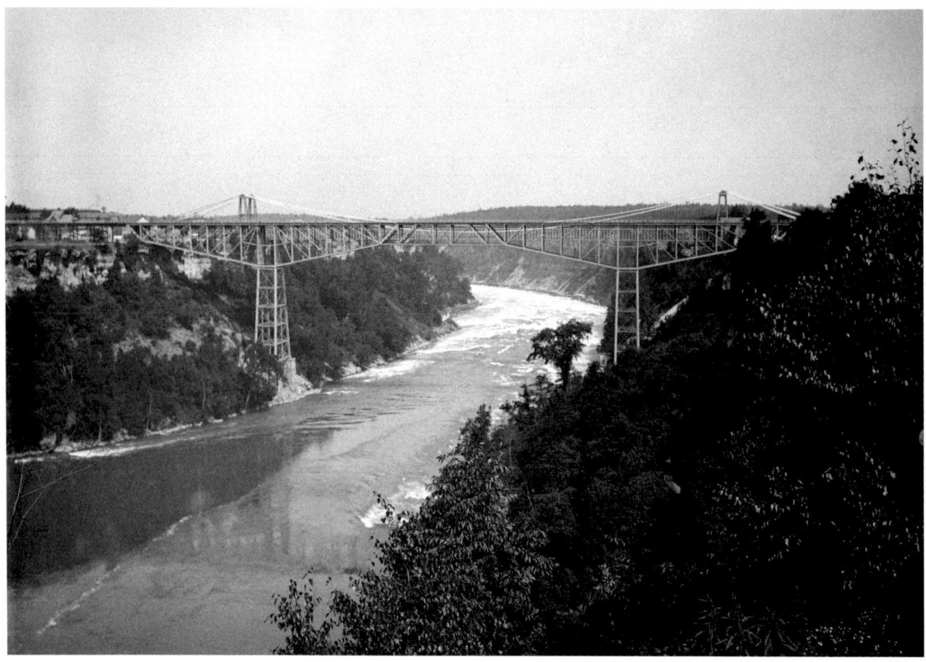

The Niagara Cantilever Bridge with the Niagara Falls Suspension Bridge in the background at Whirlpool Rapids, in a photo dating between 1880 and 1897 (when the suspension bridge was replaced). (*Library of Congress*)

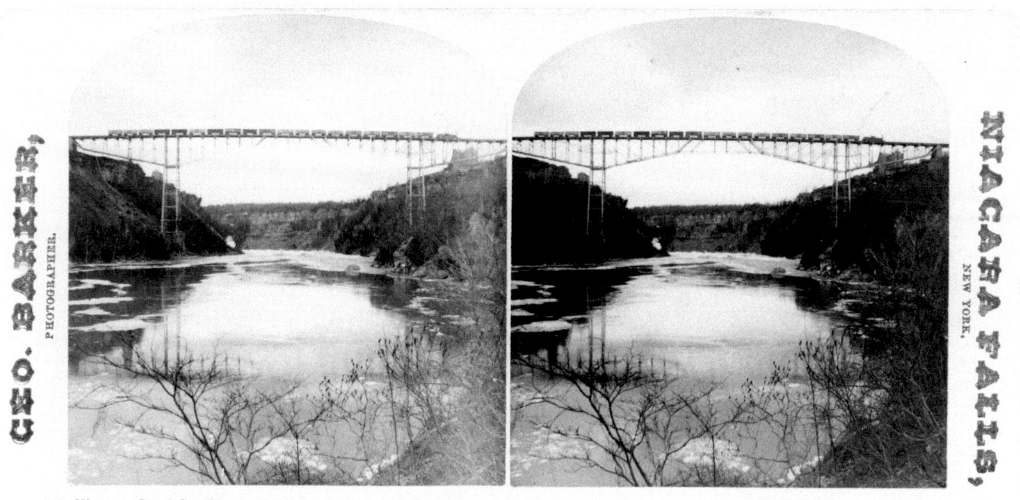

The Whirlpool Rapids Cantilever Railroad Bridge, 1890. (*Library of Congress*)

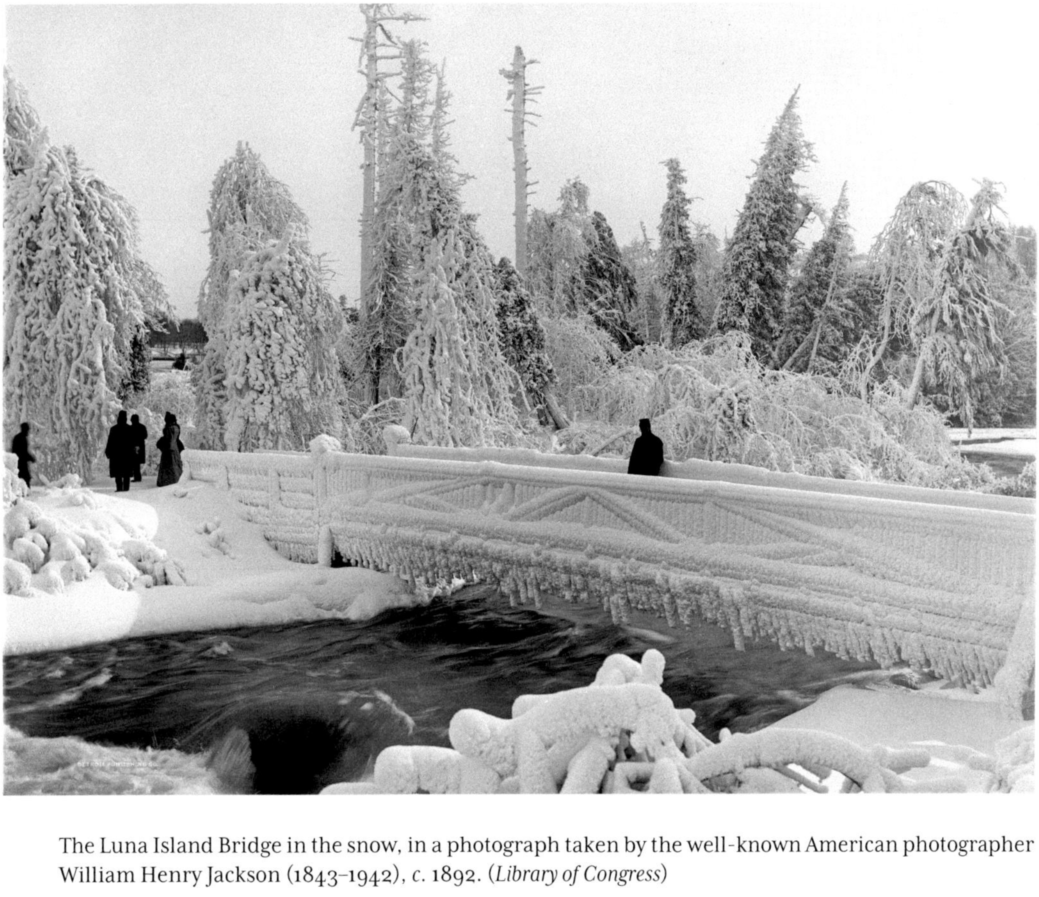

The Luna Island Bridge in the snow, in a photograph taken by the well-known American photographer William Henry Jackson (1843–1942), c. 1892. (*Library of Congress*)

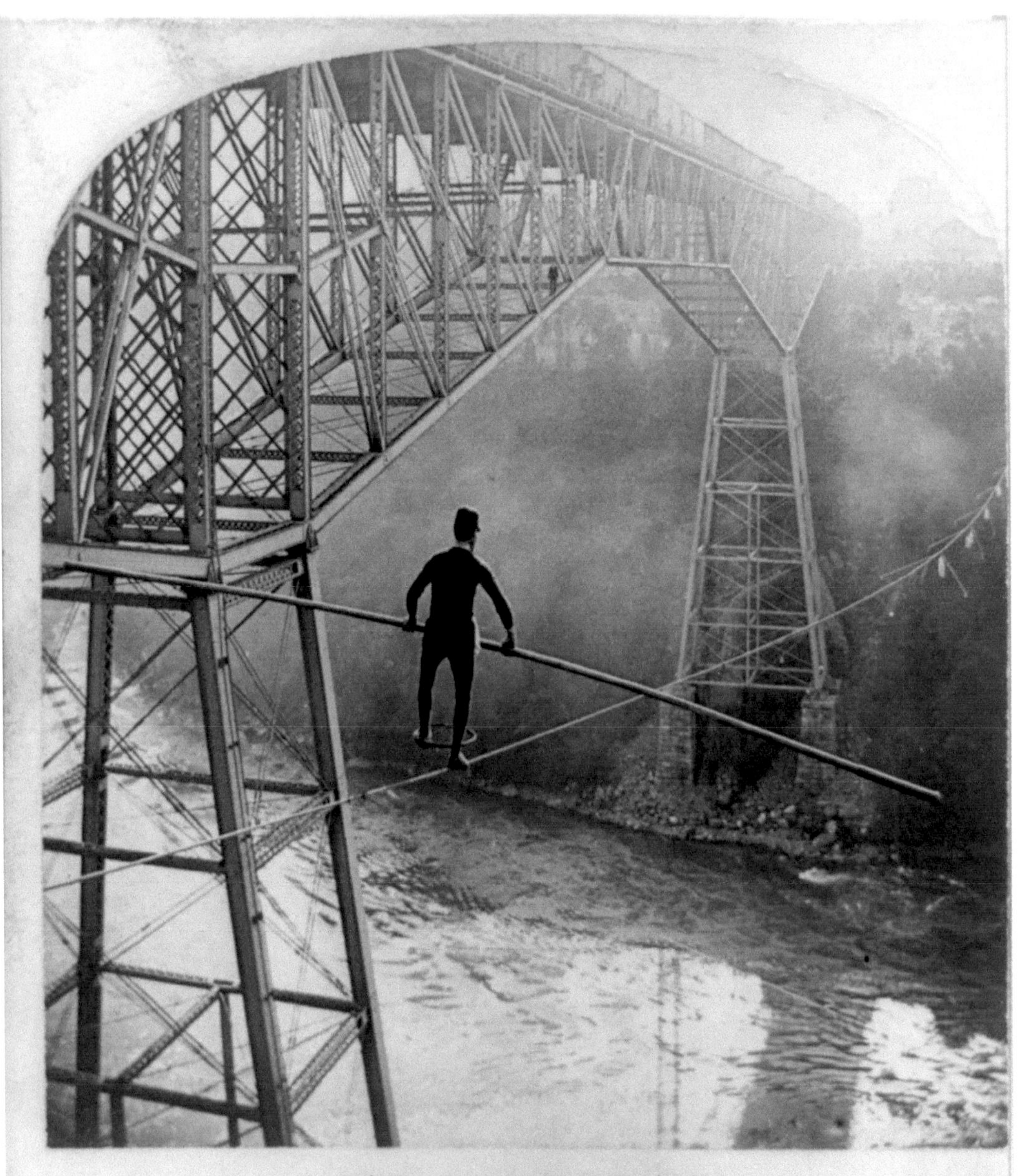

An 1890 view of Samuel John Dixon (1852–1891) crossing the river on a 7/8-inch wire. He successfully crossed the river twice, once in 1890 and once in 1891, but drowned in a lake in Ontario, Canada, later in 1891. (*Library of Congress*)

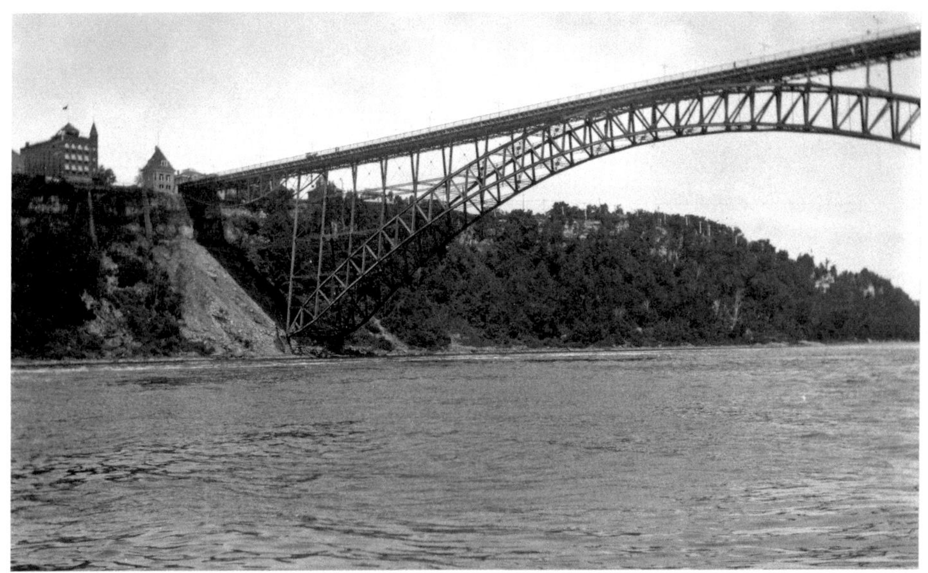

This 1895 USGS photograph's caption reads: "New carriage and electric railroad bridge, viewed from the American side, showing the character of the water surface and also the maturity of the talus slope, as indicated by forest growth. Circa 1895." (*US Geologic Survey*)

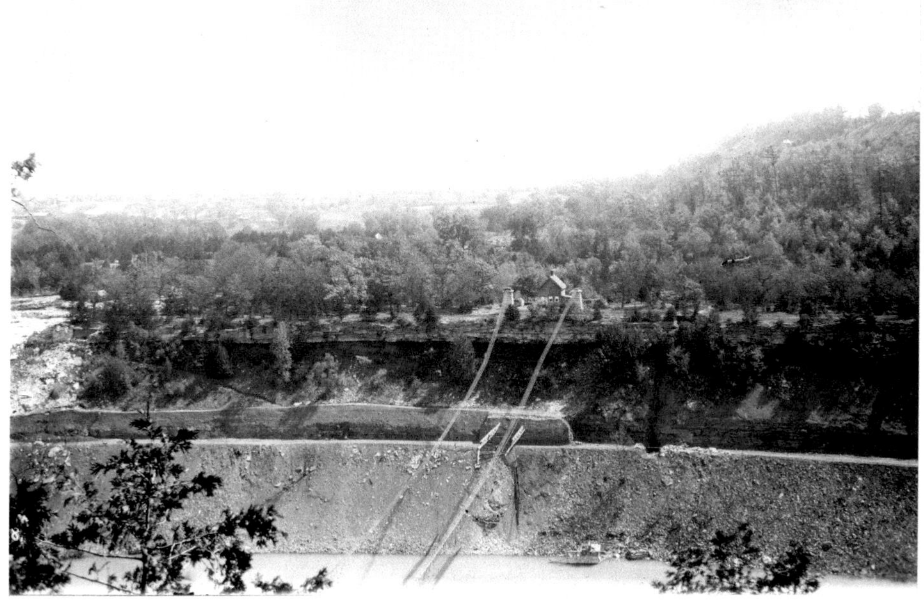

Another USGS image from 1895. "Looking eastward from Queenston Heights, Canada. The Niagara River, in the foreground, is spanned by the ruins of an old suspension bridge. The piers of this bridge stand on the quartzose sandrock of the Medina Formation. On the slope below is the newly prepared grade of the Gorge Railroad. In the distance on the left is Lewiston; on the right, the Niagara escarpment." (*US Geologic Survey*)

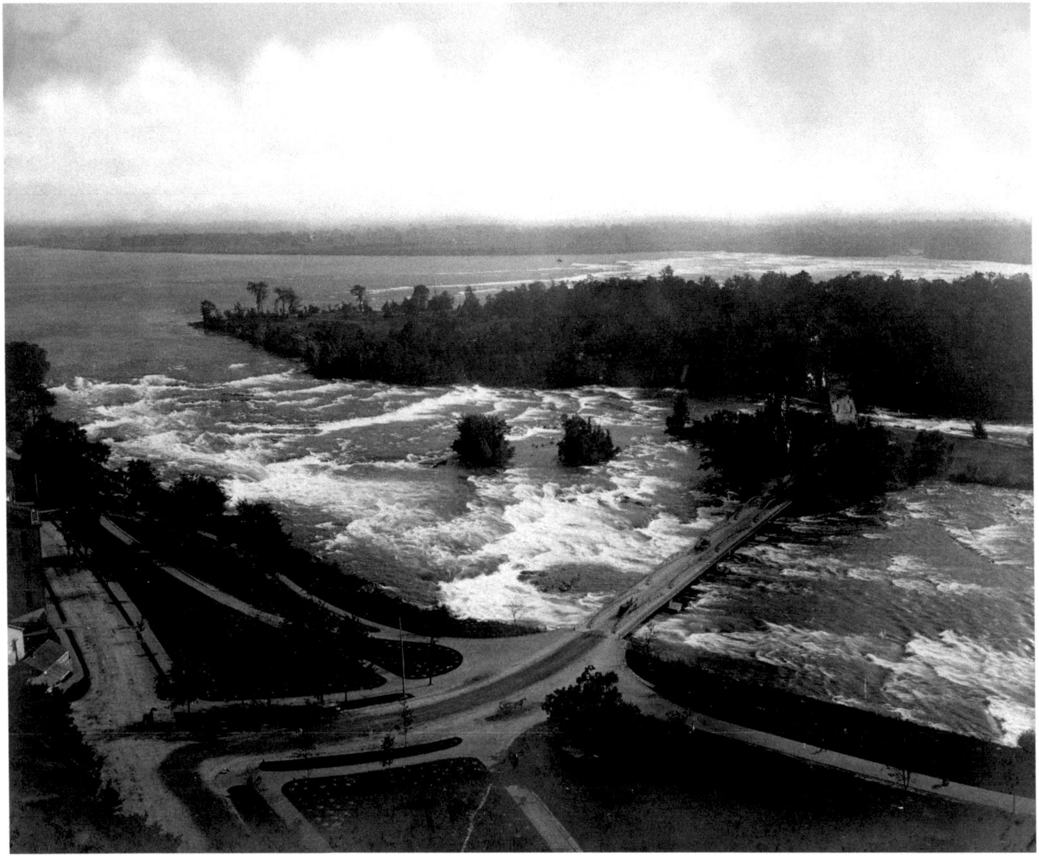

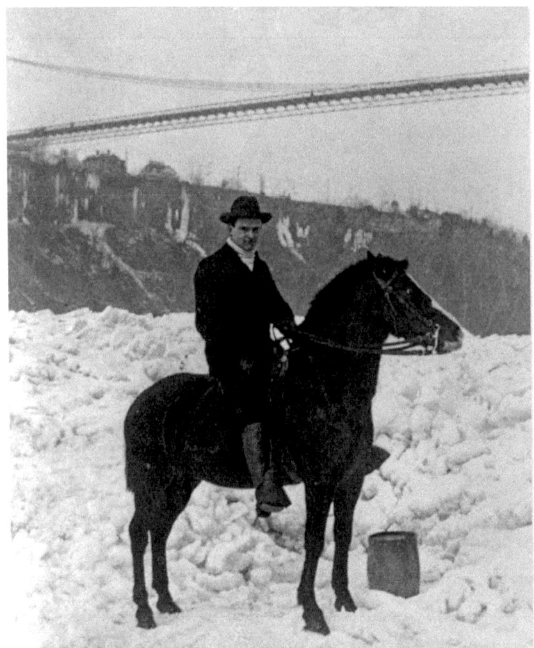

Above: The head of Goat Island and the American Rapids, as viewed from the Tower Hotel, *c.* 1896. (*US Geologic Survey*)

Left: Horseback riding over the ice bridge at the falls, 1896. (*Library of Congress*)

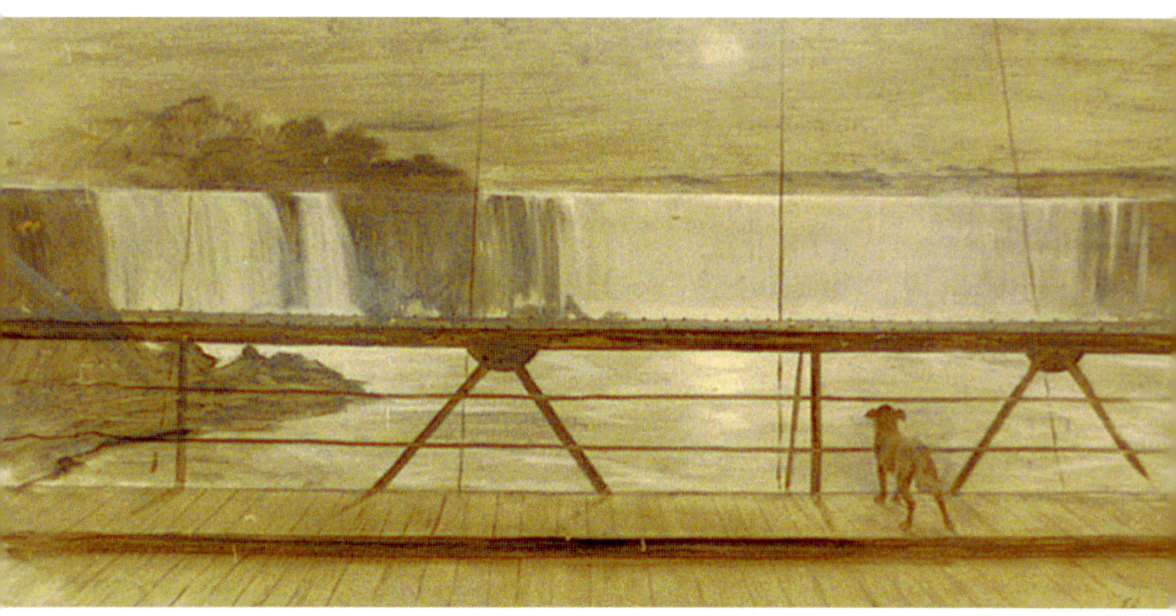

The falls from the Suspension Bridge, an oil painting by J. André Castaigne, 1896. (*Library of Congress*)

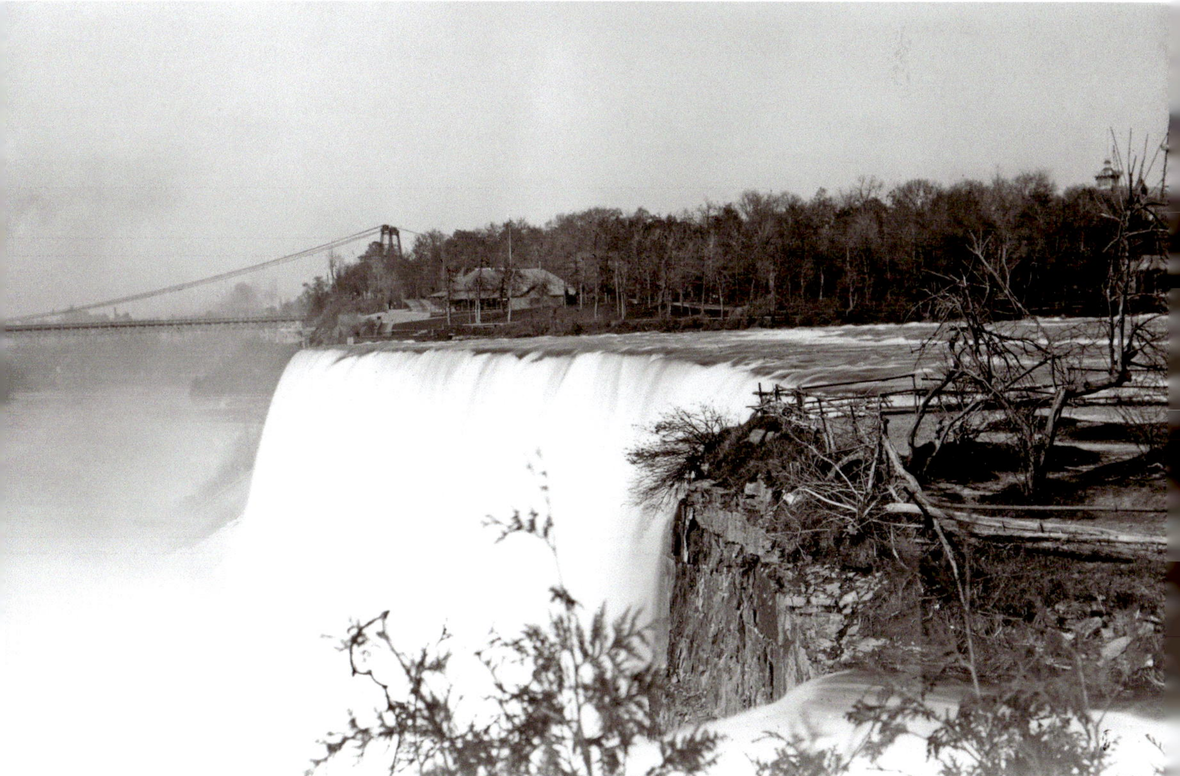

The American Falls, viewed from Goat Island in 1895. (*US Geologic Survey*)

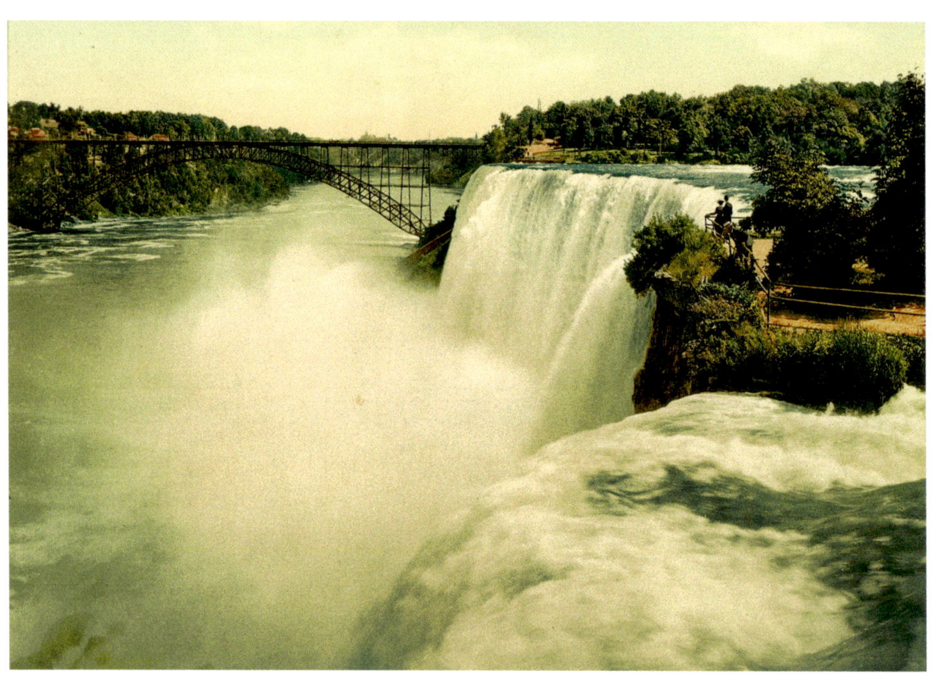

A colorized view of the American Falls and the Steel Arch Bridge from Goat Island, 1898. (*US Geologic Survey*)

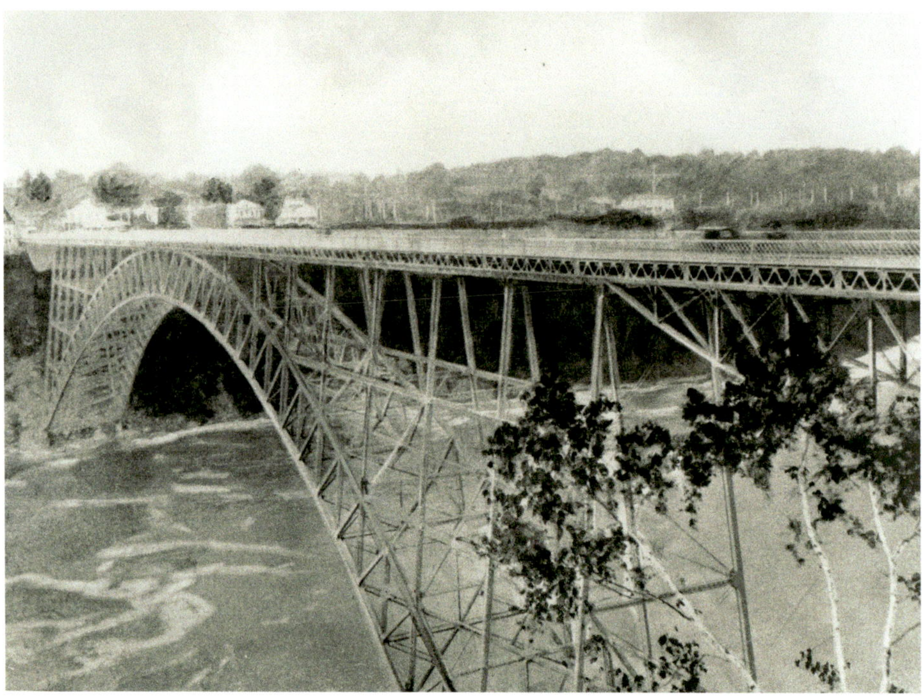

The Niagara-Clifton Bridge, also known as the Whirlpool Rapids Bridge or the Honeymoon Bridge, was designed by Leffert L. Buck and completed in 1898. (*Library of Congress*)

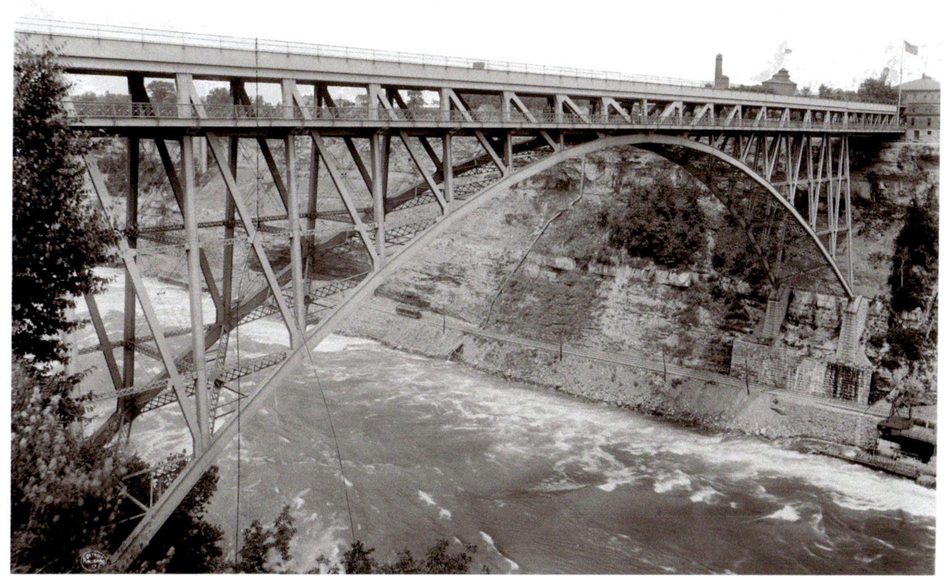

The Whirlpool Rapids Bridge was a double-track, single-arch steel railroad and pedestrian bridge, seen here *c.* 1900. (*Library of Congress*)

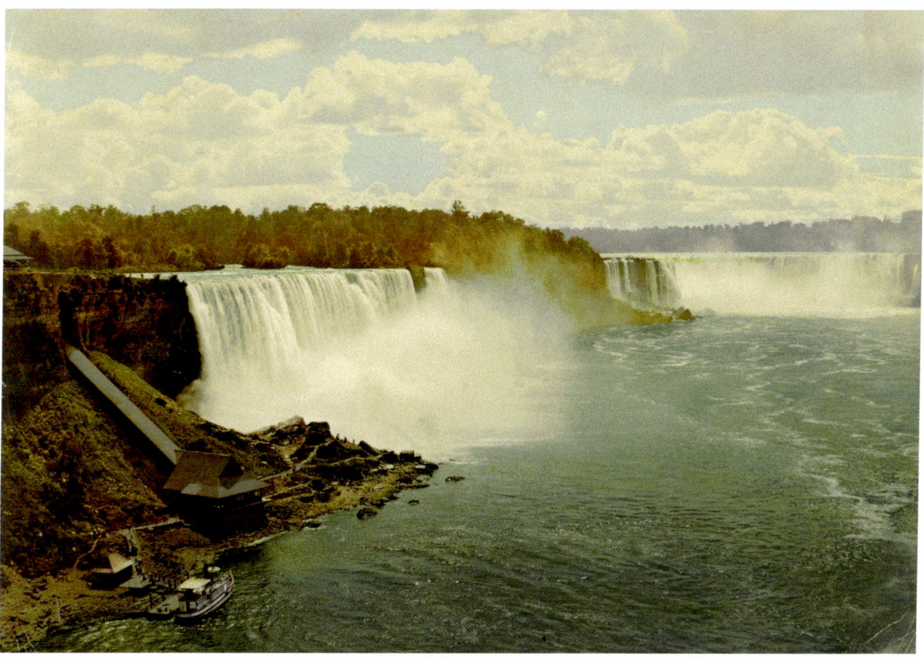

A colorized photo of Niagara Falls taken from the Steel Arch Bridge, *c.* 1900, by photographer William Henry Jackson. The *Maid of the Mist* steamboat is visible at bottom left of the image. (*Library of Congress*)

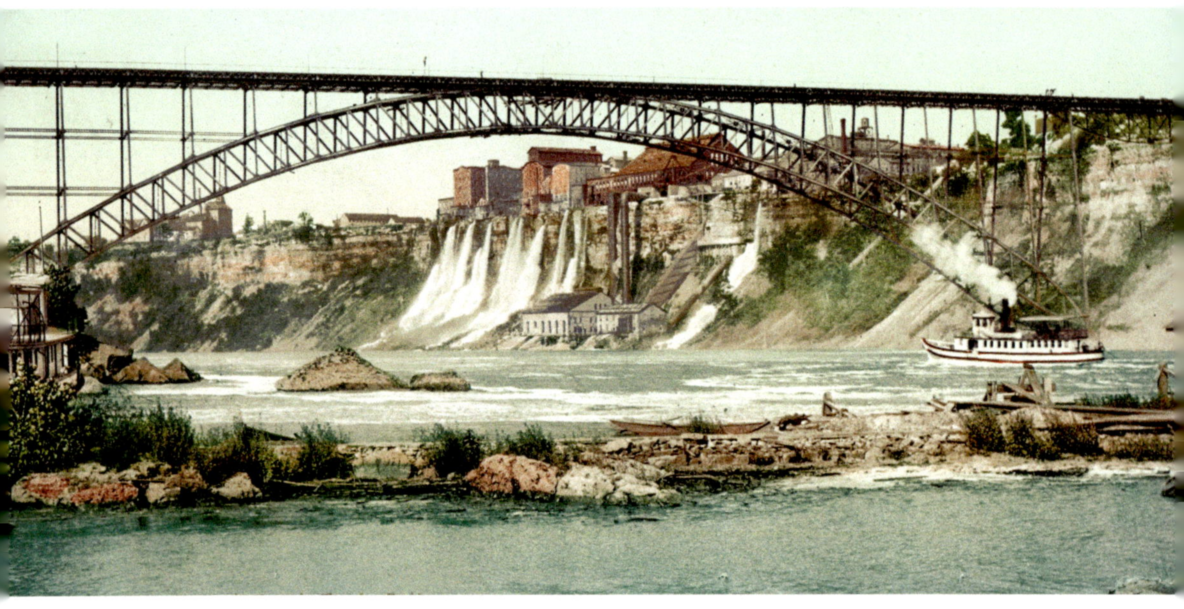

A colorized *c.* 1900 image showing mills and the Grand Trunk, aka (Upper) Steel Arch Bridge. (*Library of Congress*)

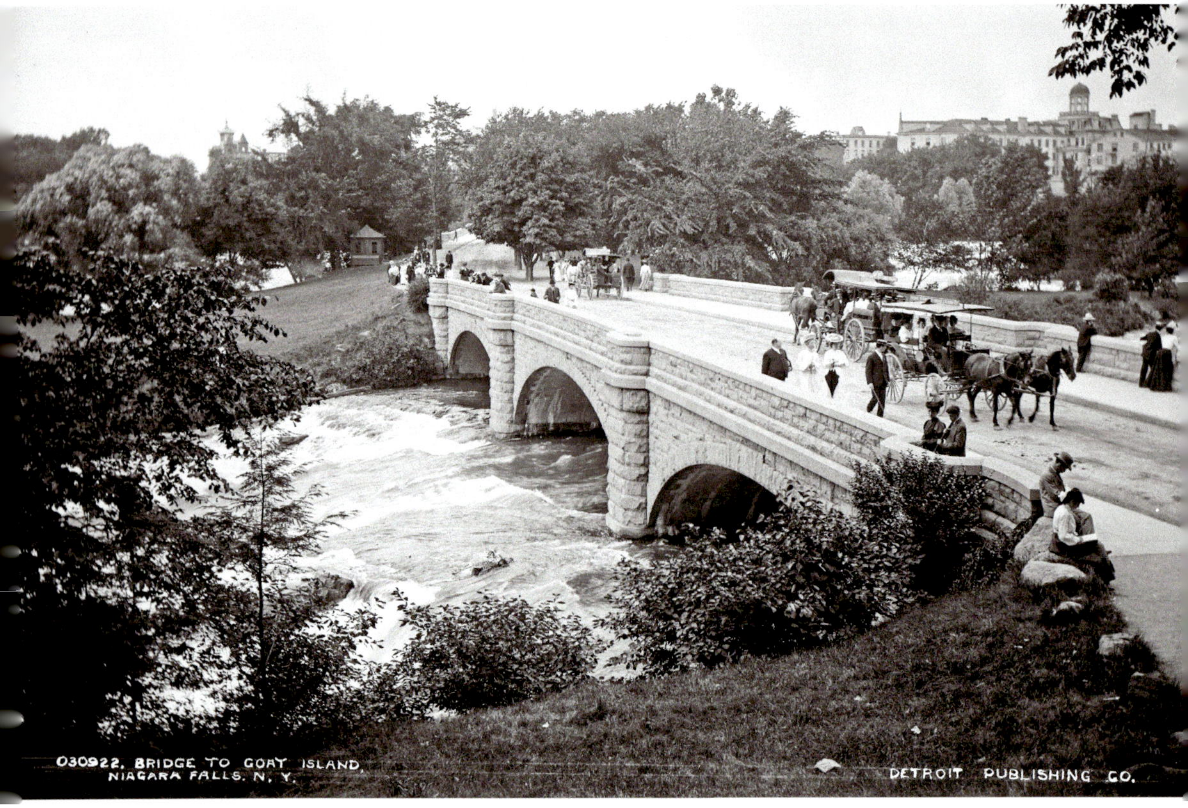

The Luna Island Bridge *c.* 1910, with pedestrians and horse and carriages crossing. (*Library of Congress*)

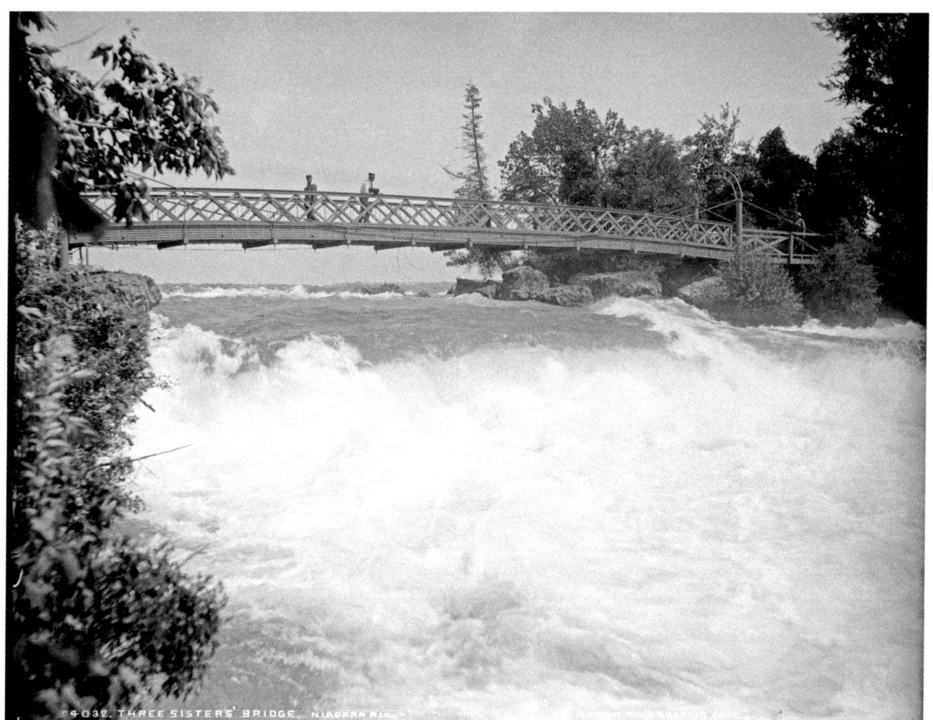

Above and below: Three Sisters Island bridge as seen in the late nineteenth century in a photo taken by William Henry Jackson, and again in a 1908 photo. (*Library of Congress*)

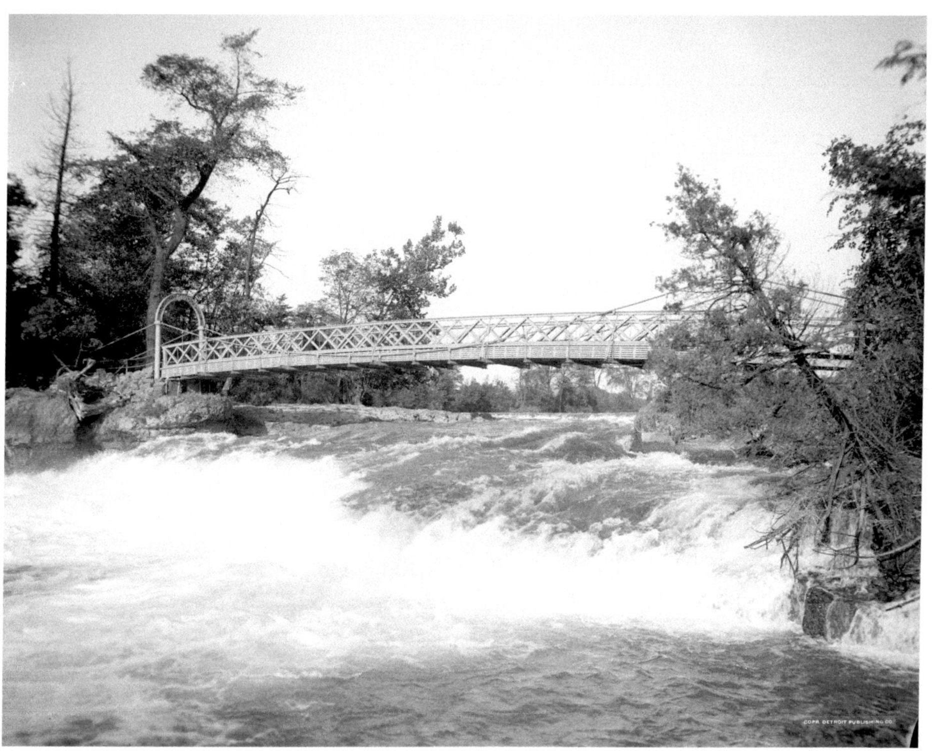

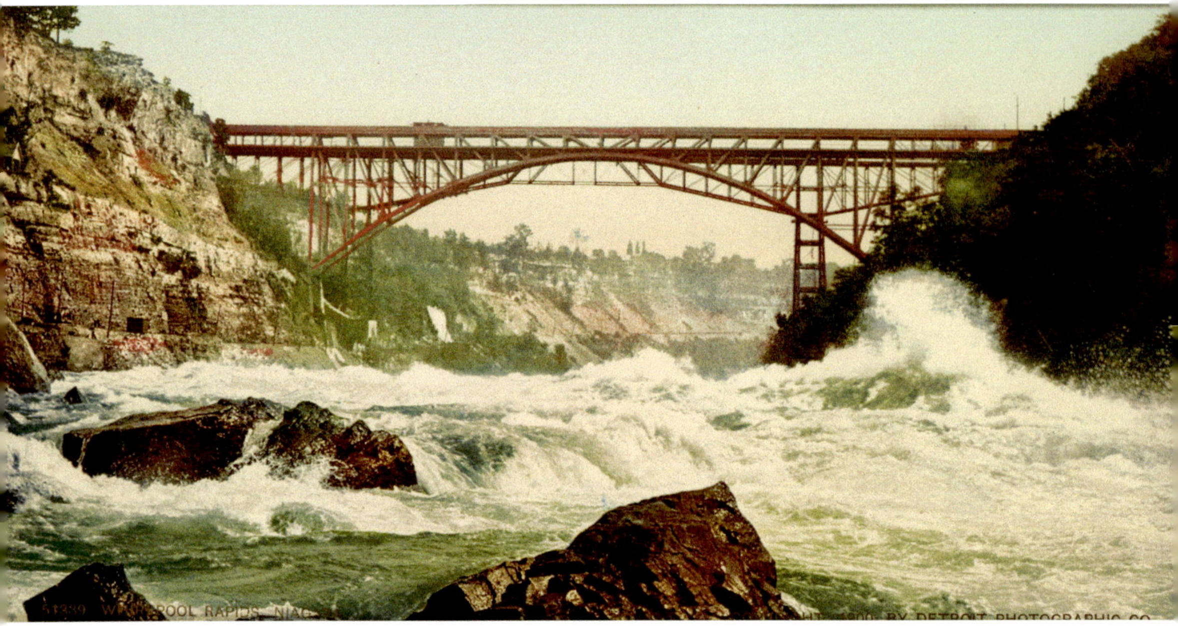

Above and below: Michigan Central Cantilever and Whirlpool Rapids (Grand Trunk Railway) bridges, c. 1900 (colorized) and 1910 (black and white). (*Library of Congress*)

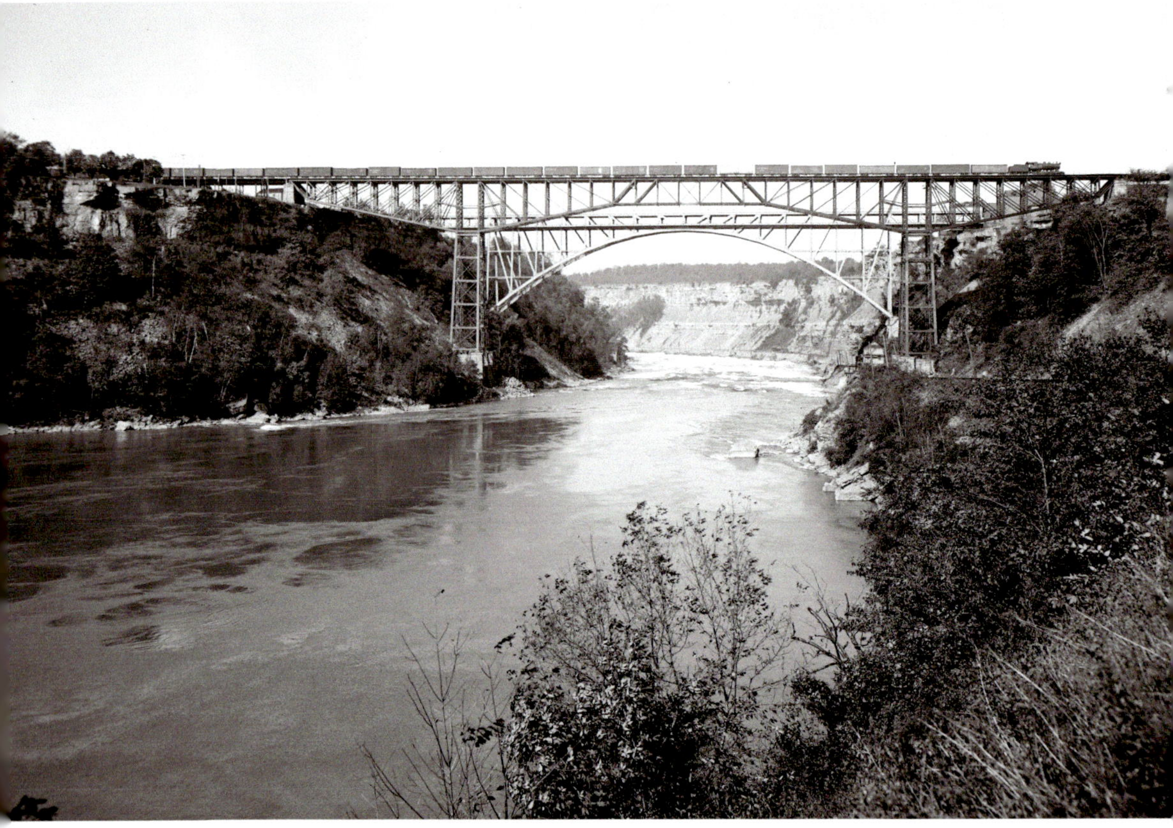

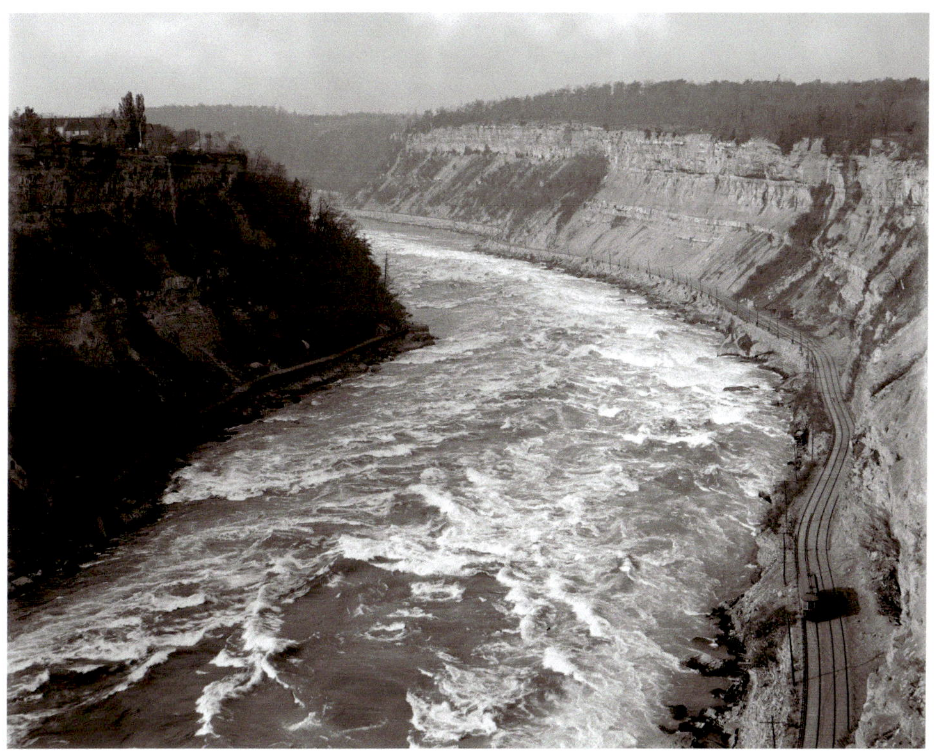

Whirlpool Rapids from the Niagara Railway Suspension Bridge, *c.* 1910. (*Library of Congress*)

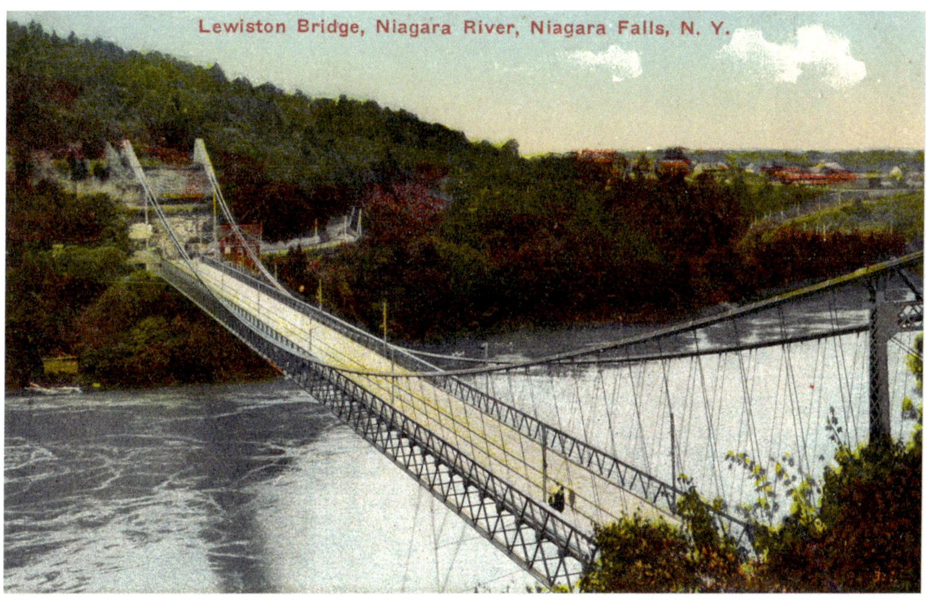

The suspension bridge from Lewiston to Queenston shown in this *c.* 1910 image was opened to the public in July 1899. It was 1,050 feet long and 80 feet above the water. The first suspension bridge on this site was built in 1850 and destroyed by a hurricane in 1865.

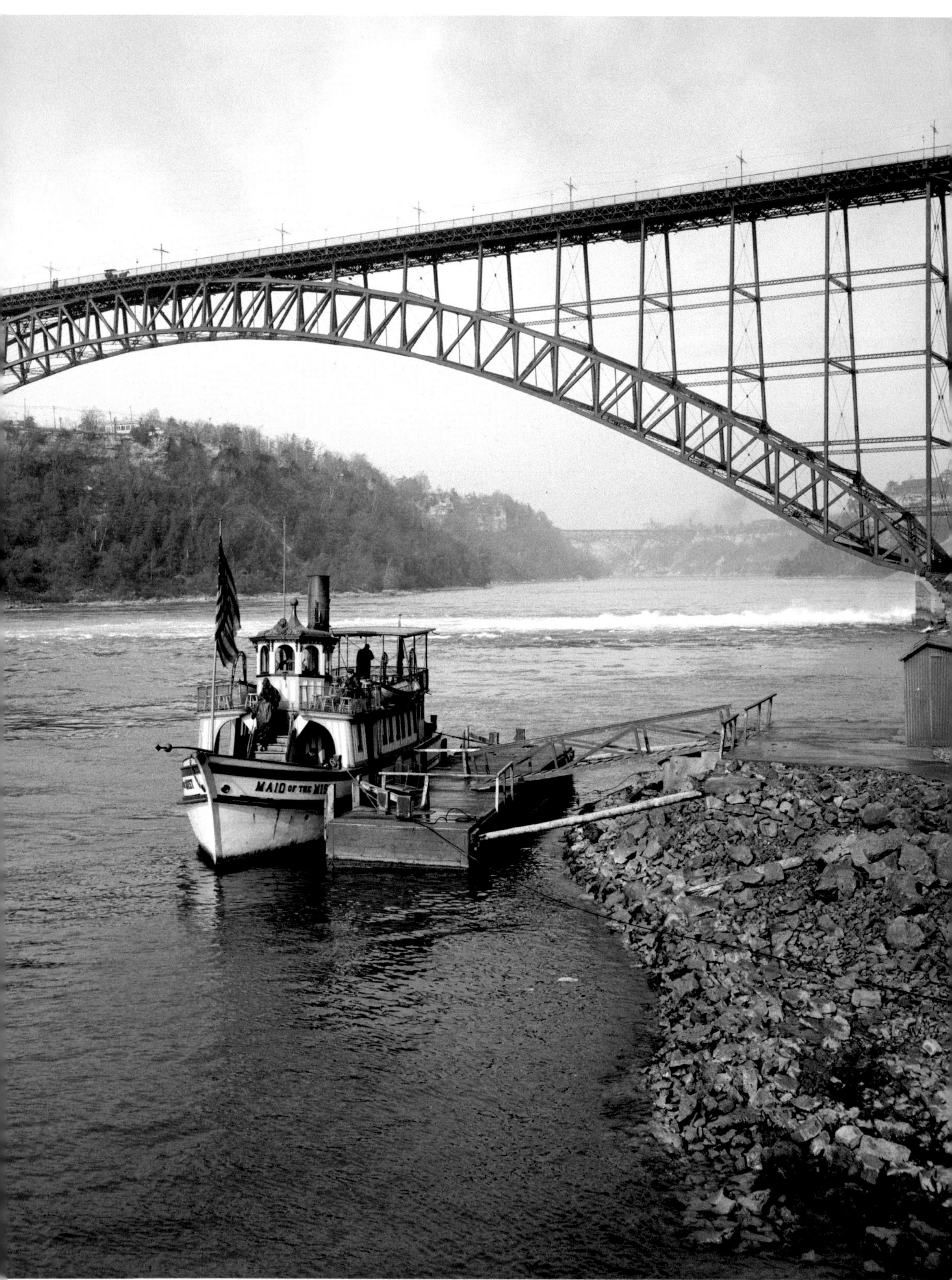

MAID OF THE MIST

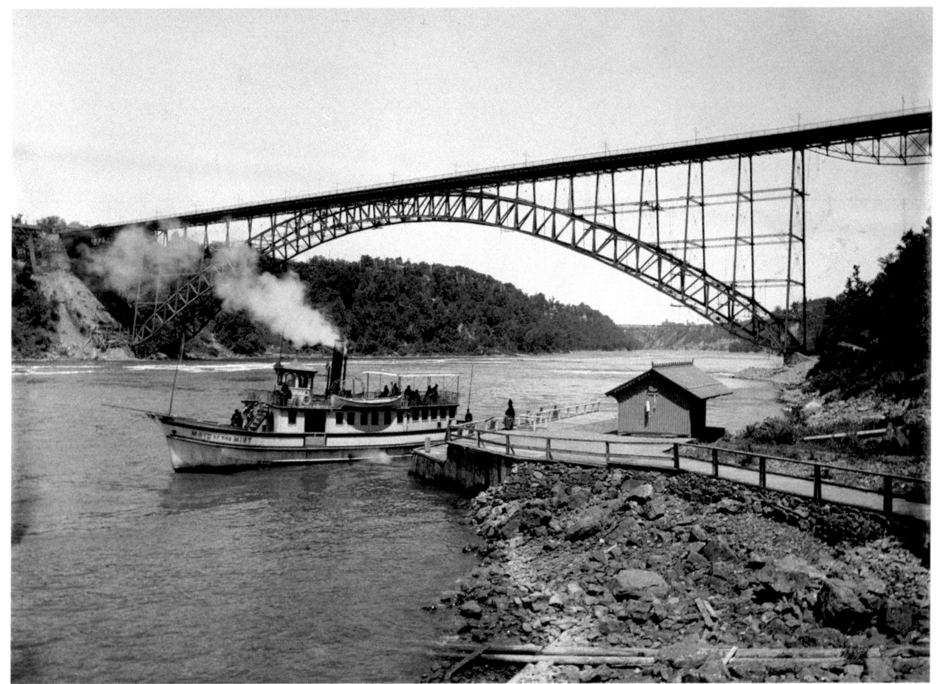

Above and below: The *Maid of the Mist* tour boat has been a Niagara Falls tradition since 1846. This trio of c1910 images show the *Maid of the Mist II*, an 89-foot-long white oak ship with double engines and a 19-foot beam with the Upper Steel Arch Bridge aka Honeymoon Bridge, built in 1897. (*Library of Congress*)

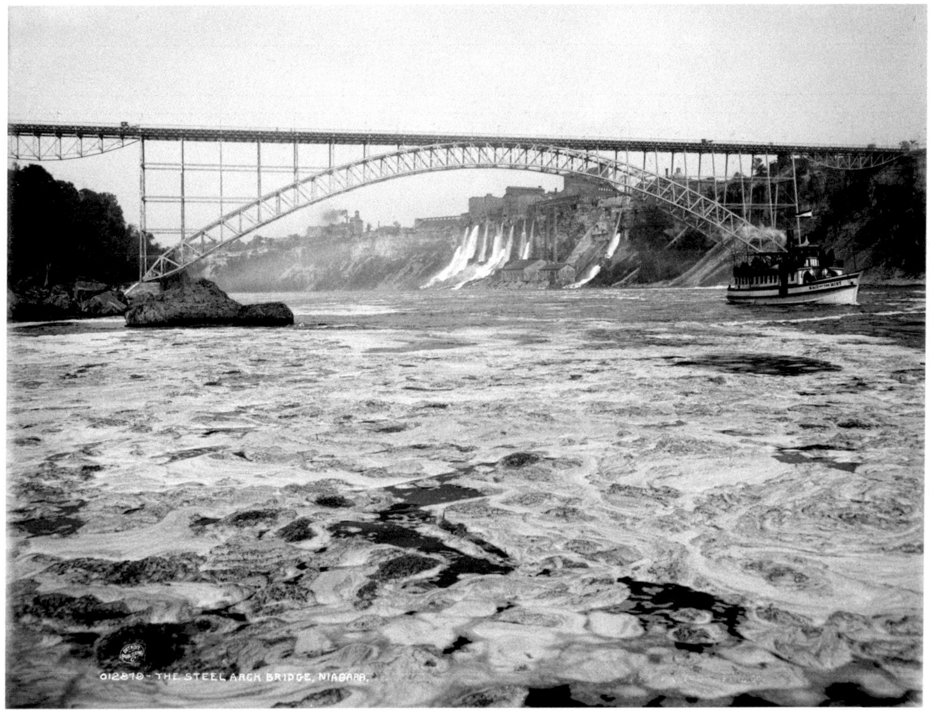

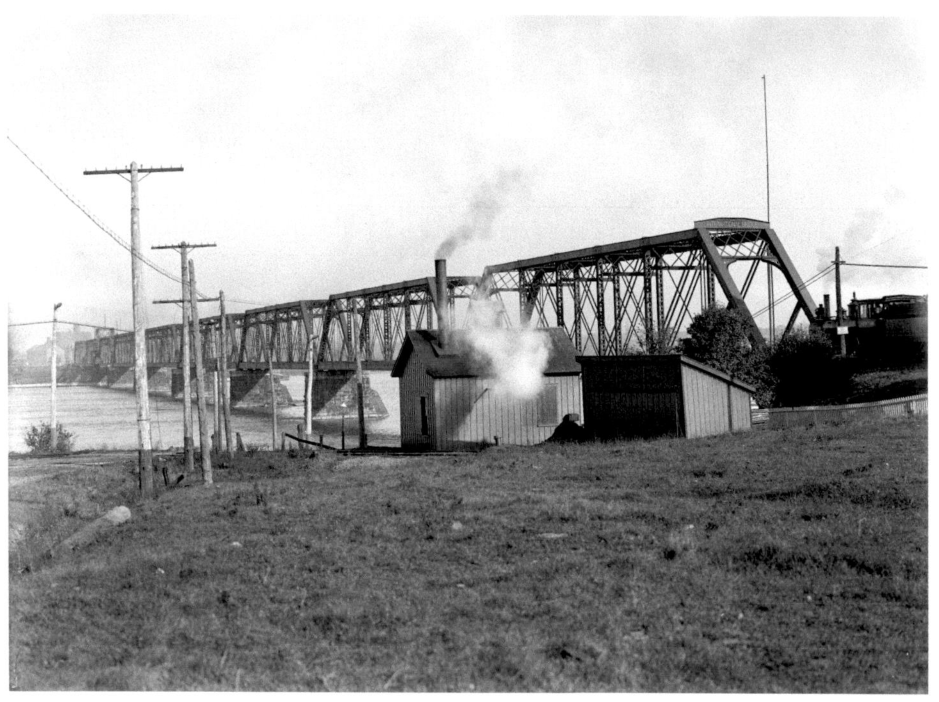

Above and below: The single-track International Railway Bridge between Buffalo, New York, and Fort Erie, Ontario, was completed in 1873. It is 3,651 feet long. The swing-span bridge was designed by Casimer S. Gzowski and D. L. MacPherson. These photos show the bridge *c.* 1910. (*Library of Congress*)

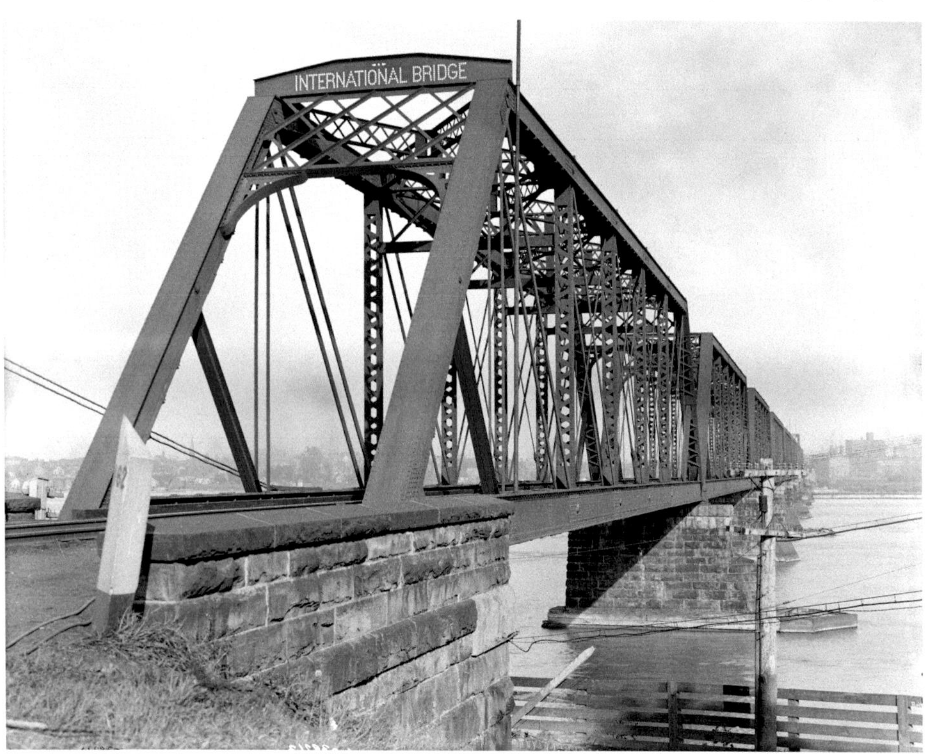

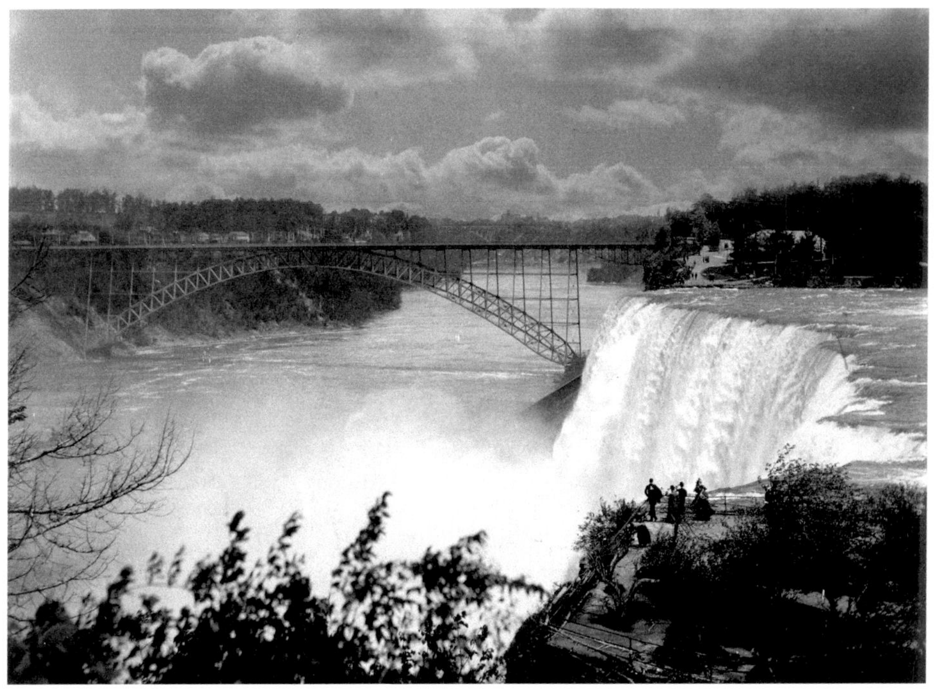

The American Falls and the International Bridge, seen from Stedman bluff on Goat Island, 1911. (*US Geologic Survey*)

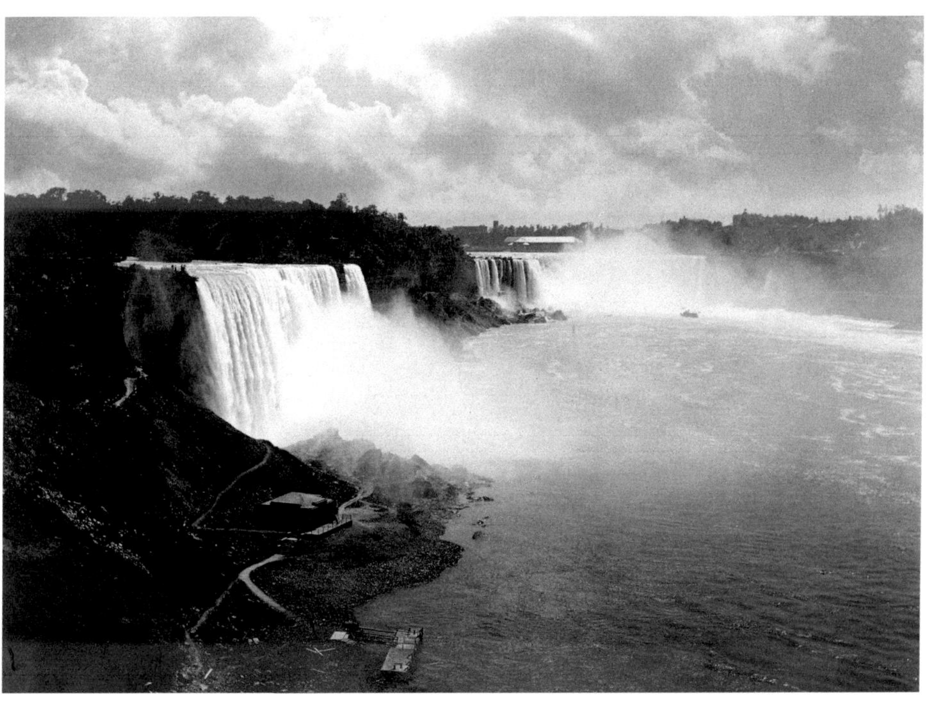

Niagara Falls, as seen from the east end of the International Bridge, 1911. (*US Geologic Survey*)

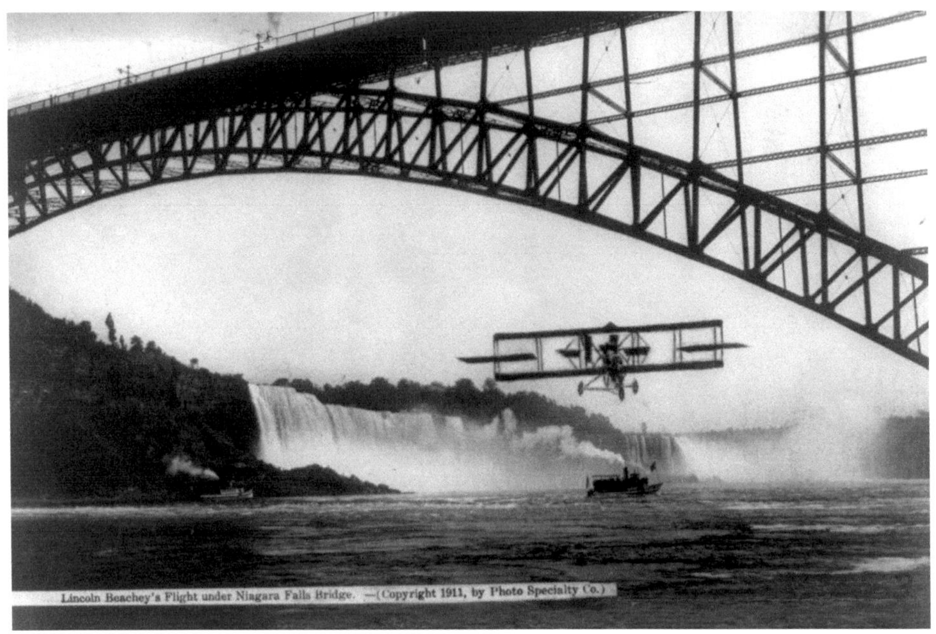

Above and below: Two images of stunt pilot Lincoln Beachey's (1887–1915) flight under the Steel Arch Bridge in June 1911. He had just become the first person to fly a plane over Niagara Falls, winning a prize of $1,000. This was an additional stunt he performed; he was only 35 feet above the water at one point. (*Library of Congress*)

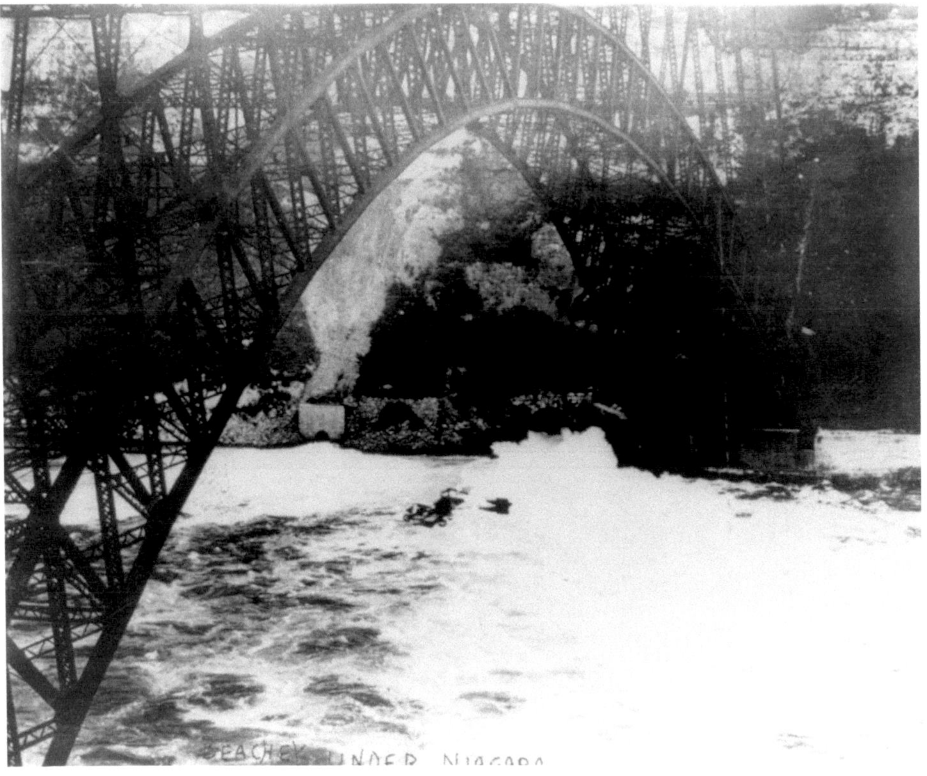

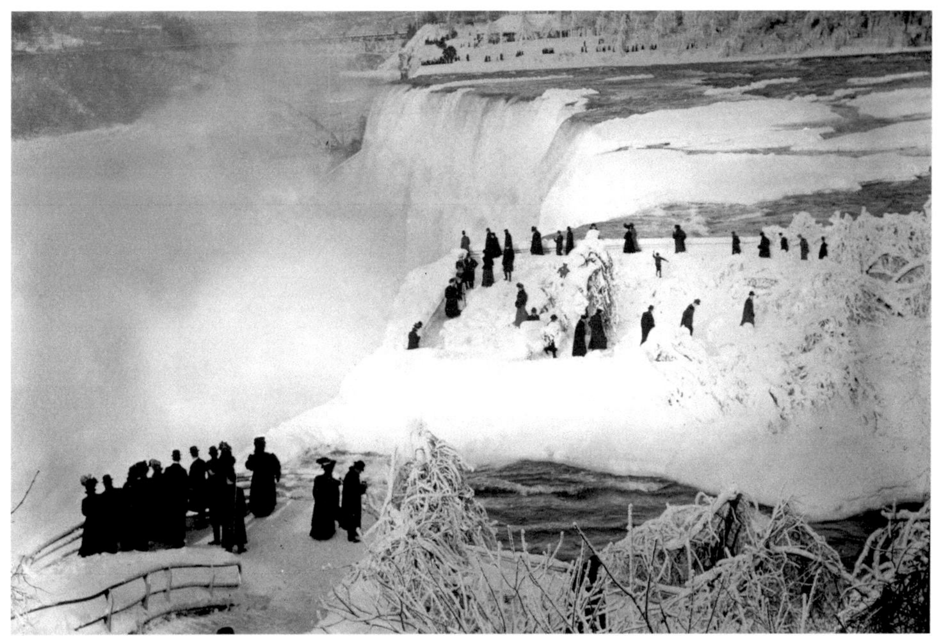

People venturing out onto an "ice bridge" at Niagara Falls in 1912. Due to its location, in winter the area often experiences extended periods of below-freezing temperatures, allowing a solid enough platform of ice to support a good deal of weight. (*Library of Congress*)

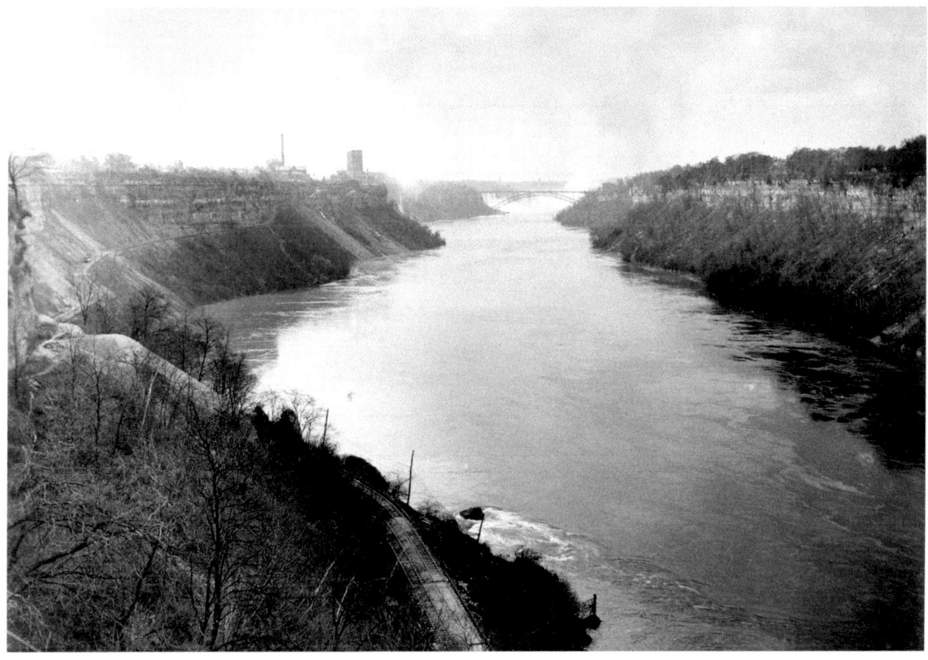

The upper great gorge of the Niagara River, looking south from the west end of the cantilever railroad bridge; the falls and International Bridge are in distance nearly two miles away, 1913. (*US Geologic Survey*)

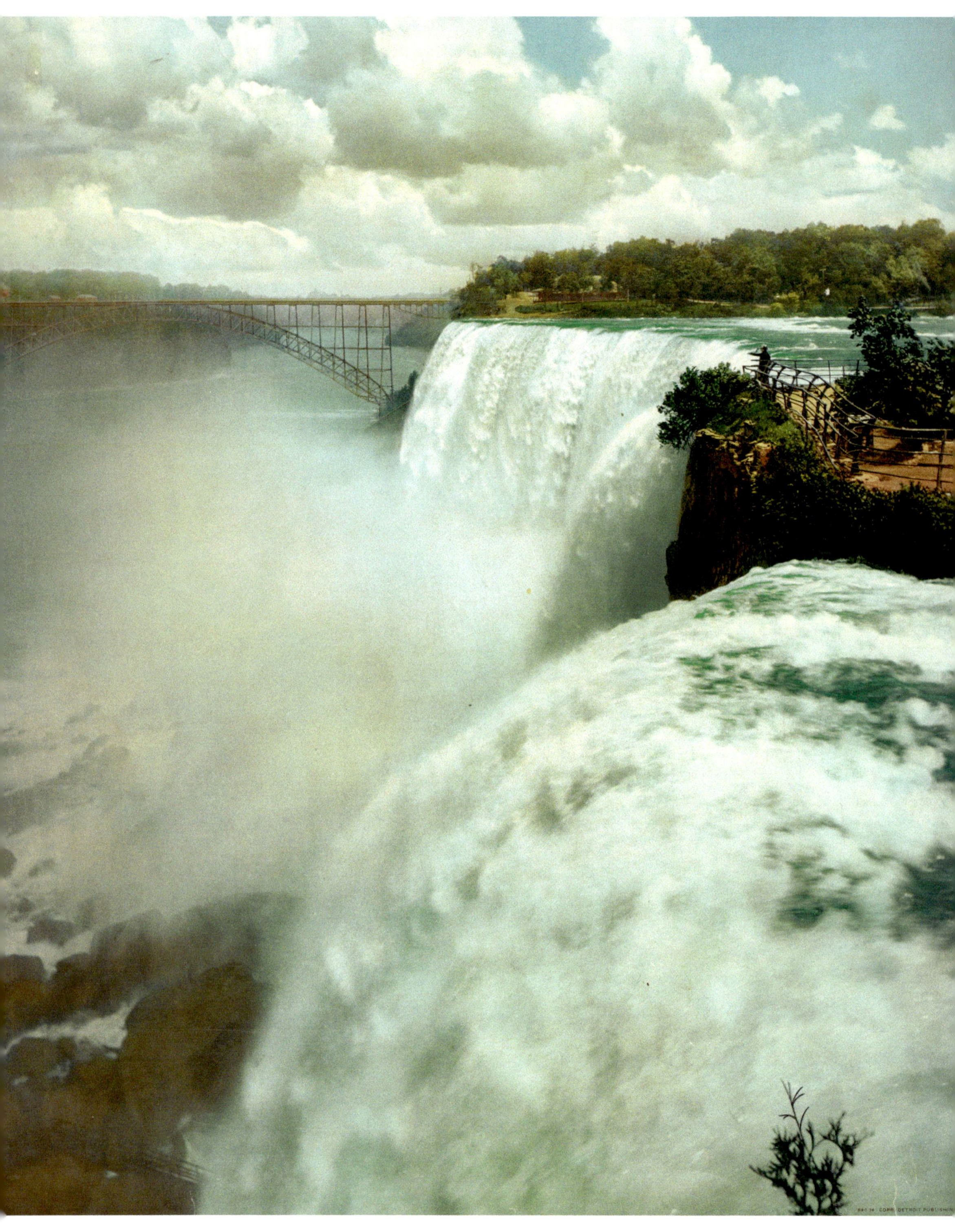

The American Falls from Goat Island, with the Steel Bridge in the background, 1912, photographed by William Henry Jackson. (*Library of Congress*)

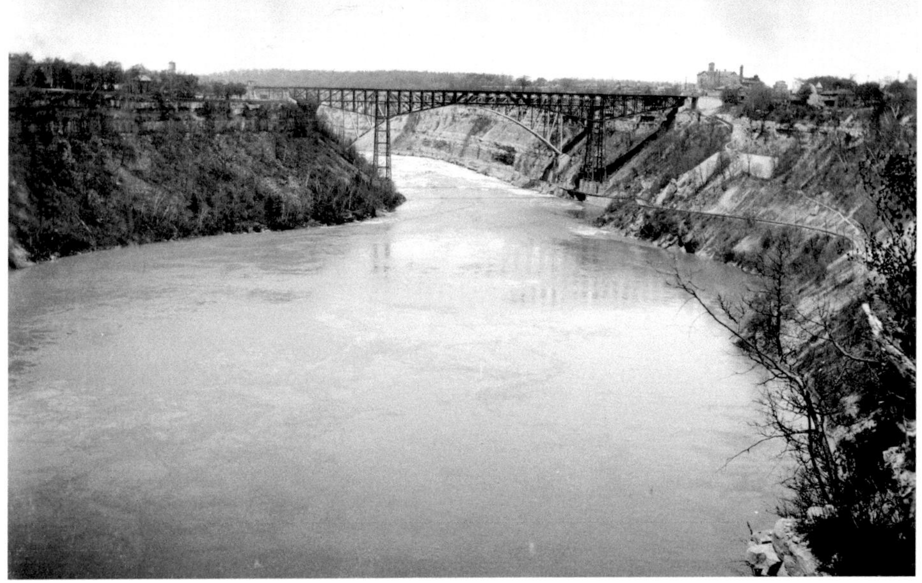

This 1913 USGS photo shows the lower end of the upper great gorge in the foreground and the head of the narrow Whirlpool Rapids gorge in the middle distance, looking north under railroad bridges; note the contrast in width of the gorge and behavior of the water in the two sections. (*US Geologic Survey*)

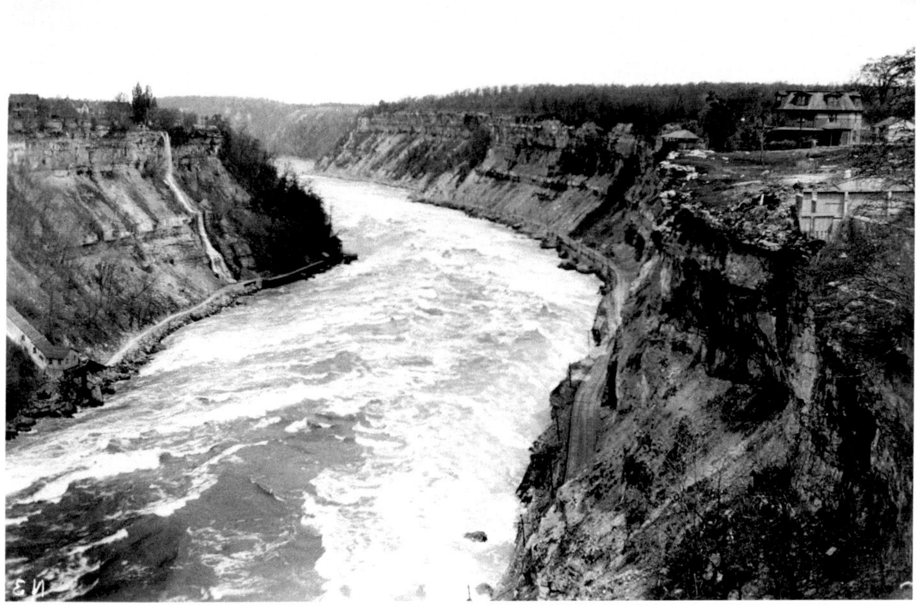

Whirlpool Rapids gorge, looking north (downstream) from the near east end of the Grand Trunk Railway Bridge in 1913. (*US Geologic Survey*)

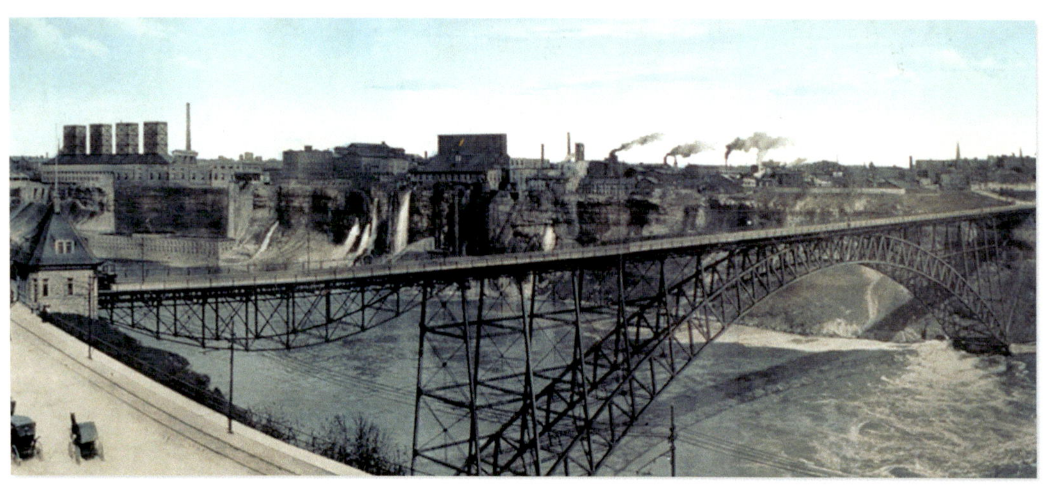

A view of the bridge by David Ellis of Buffalo, published in 1913, taken from Victoria Park, Canada (*Library of Congress*).

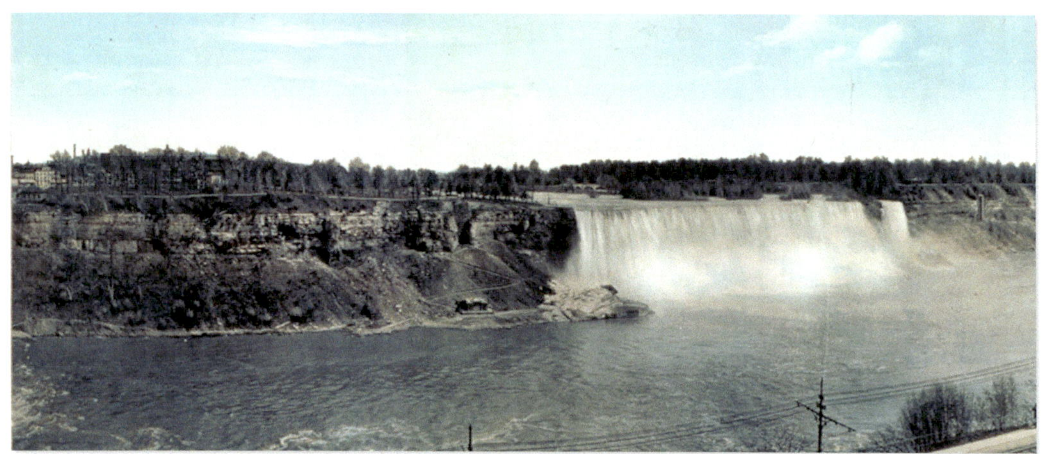

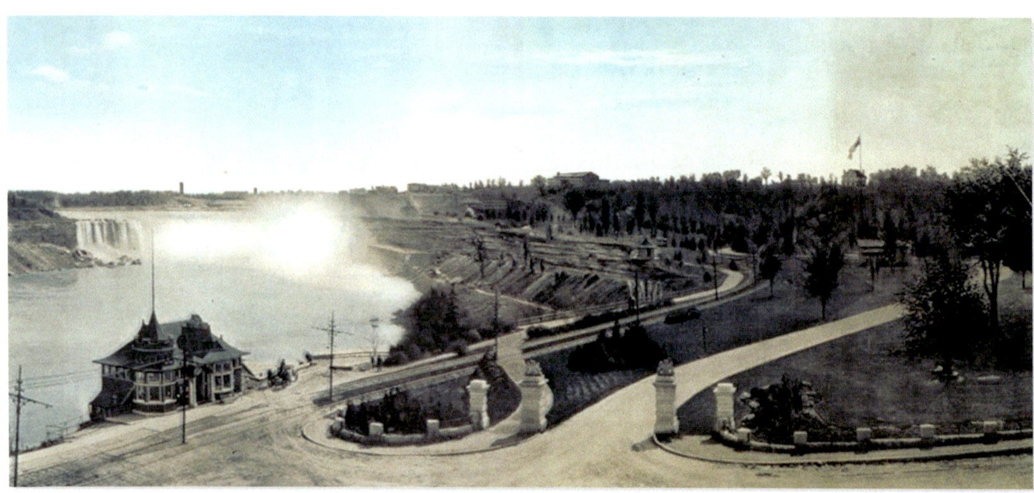

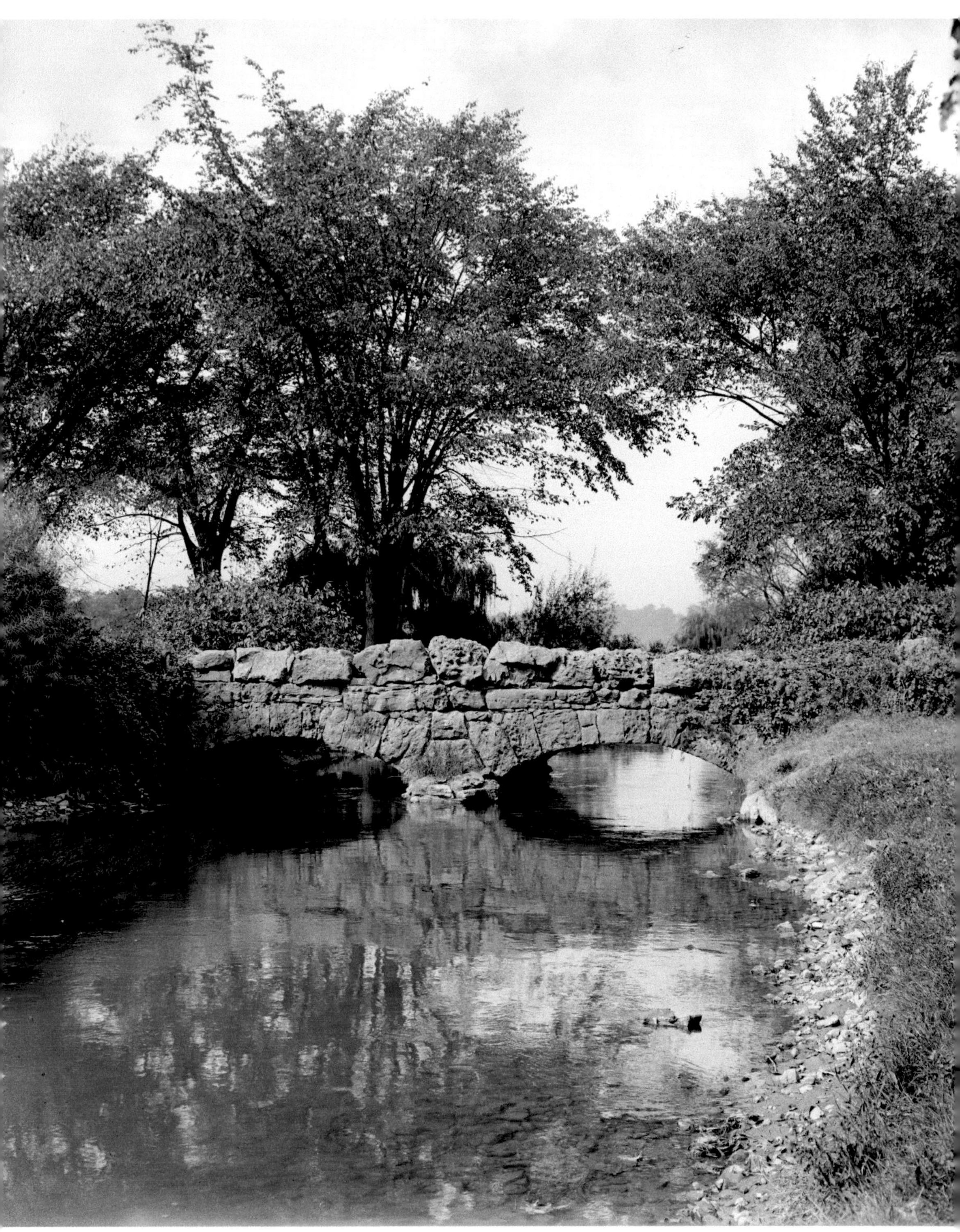

The Willow Island Bridge *c.* 1905–20. Located near Goat Island, Willow Island was created in 1759 and ceased to exist in the 1960s when the canal between it and the mainland was filled in. (*Library of Congress*)

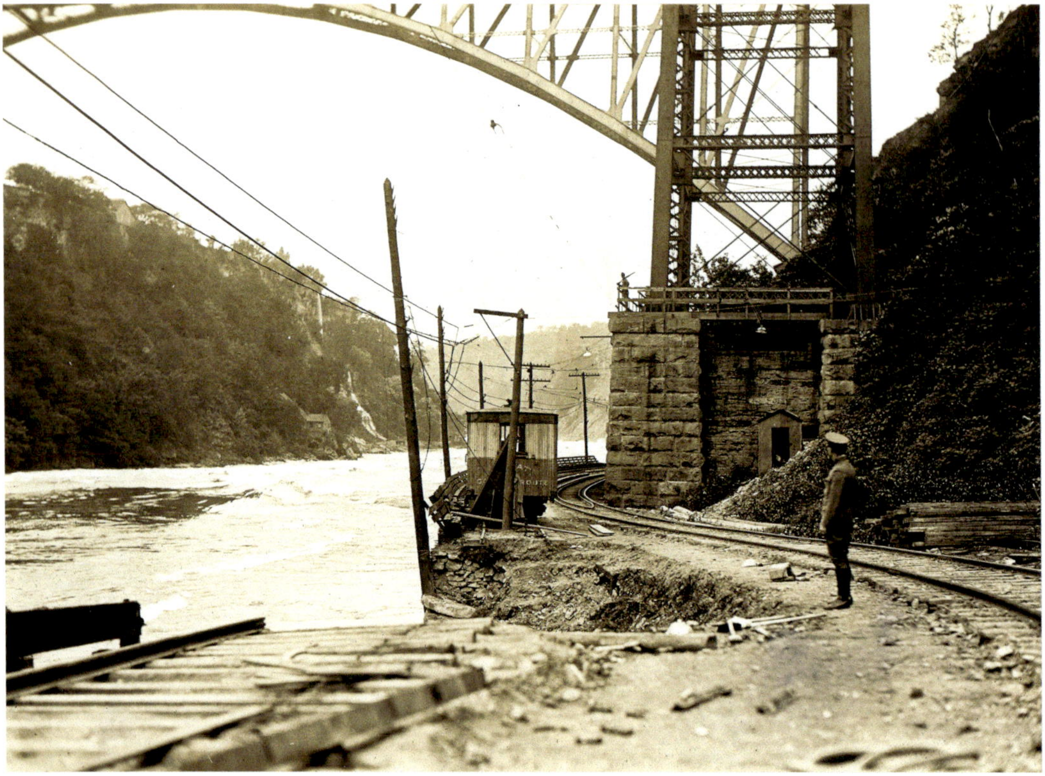

Soldiers guarded the piers of the Niagara Falls bridges during World War I. This image was taken on July 15, 1917. (*National Archives*)

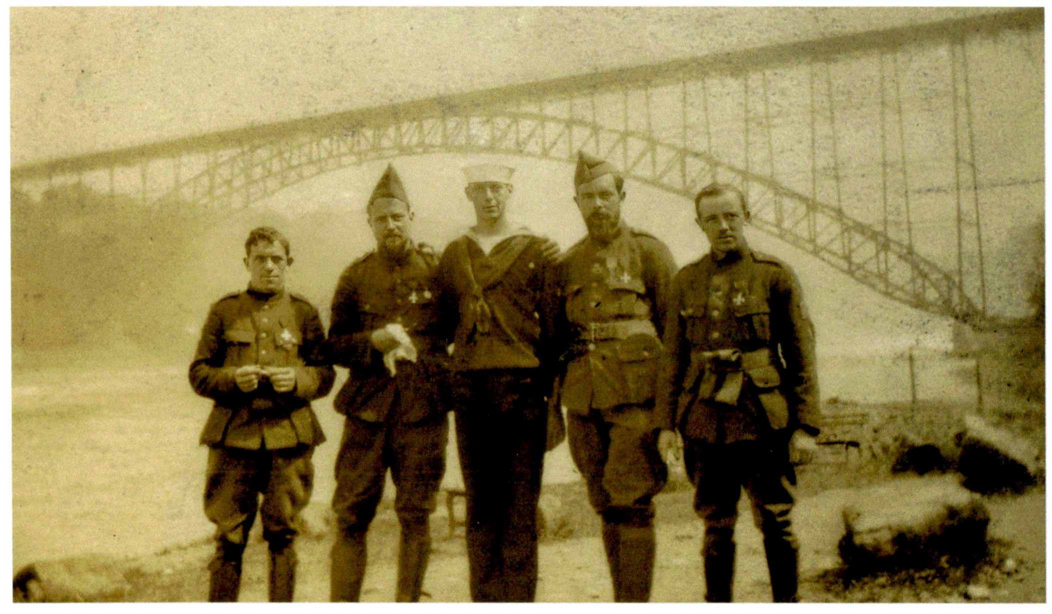

A group of Belgian World War I heroes standing near the Steel Arch Bridge in April 1918. (*National Archives*)

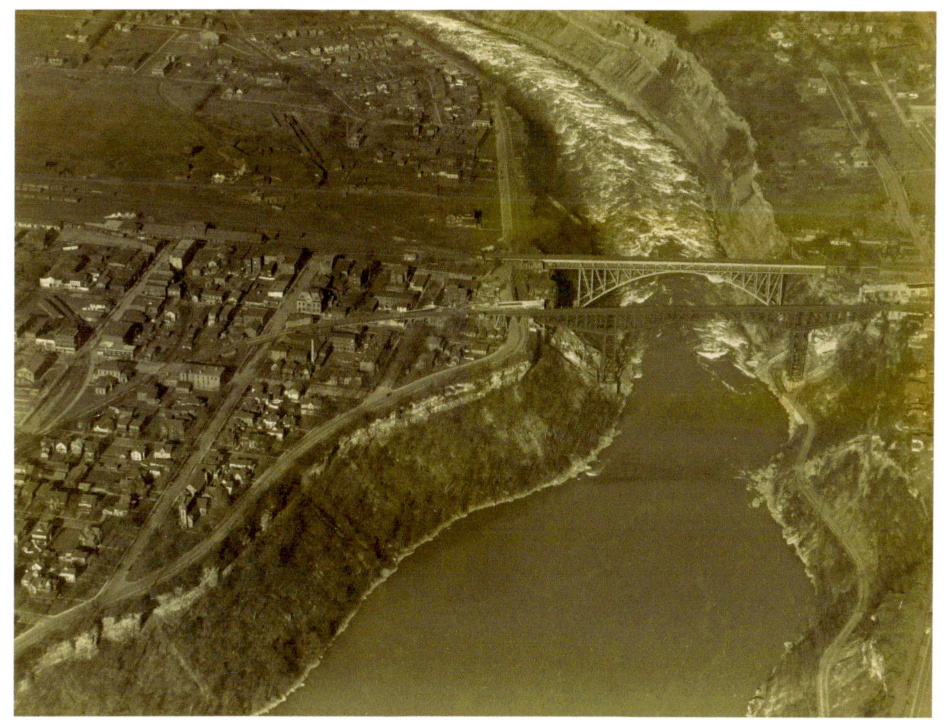

Above and below: Two 1924 aerial views of the Niagara Cantilever Railroad Bridge (1883–1925) next to the Whirlpool Rapids Bridge. The old railroad bridge was about to be replaced by the Michigan Central Railroad Bridge the following year. (*National Archives*)

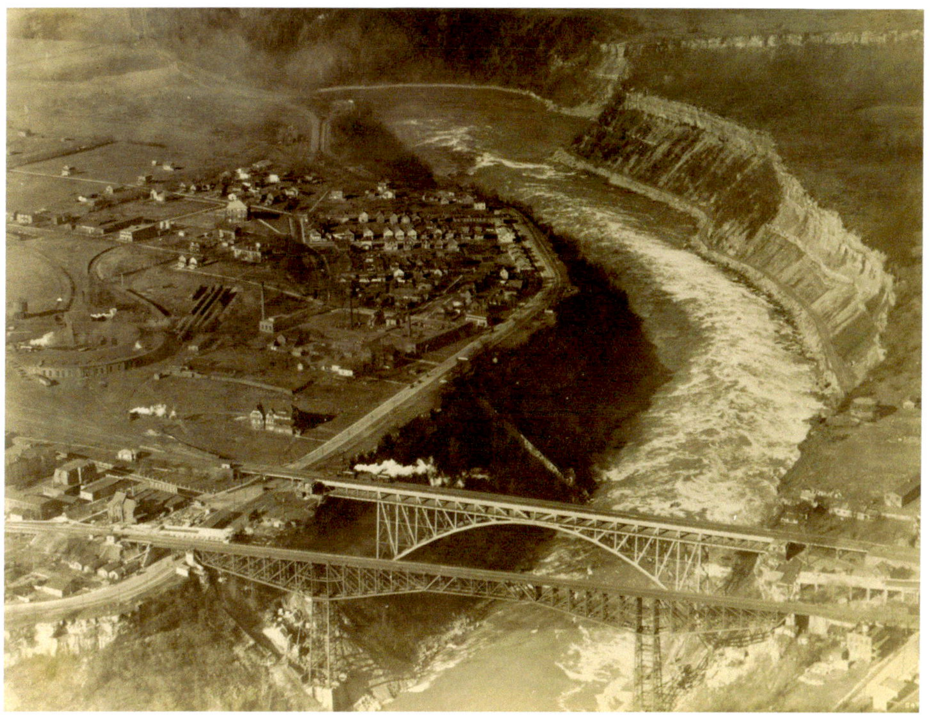

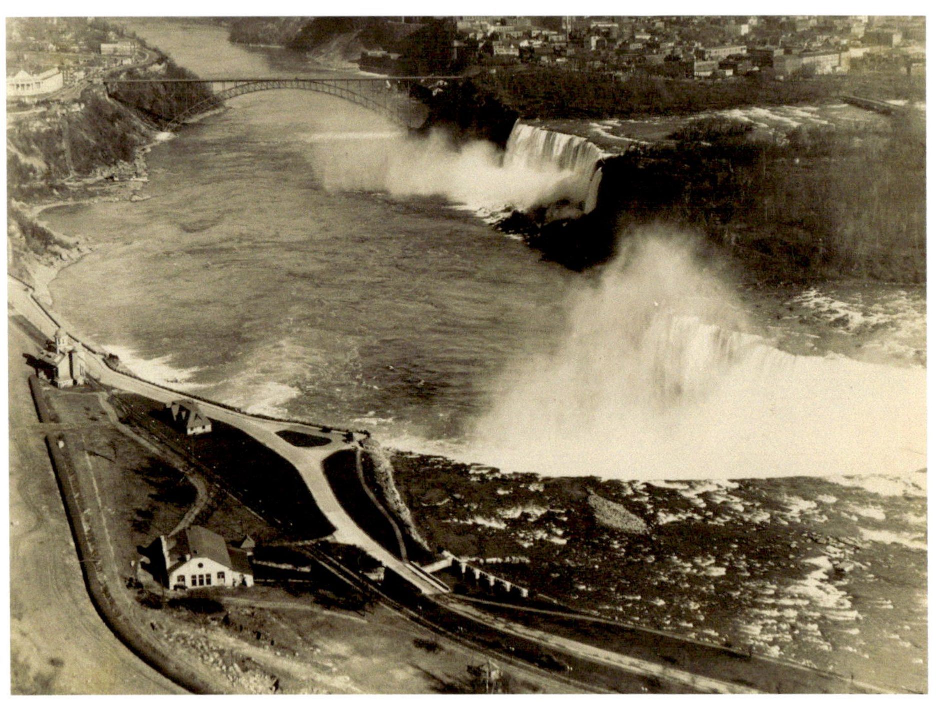

A 1924 aerial view of the falls and the steel bridge, taken by Lieutenant Stevens. (*National Archives*)

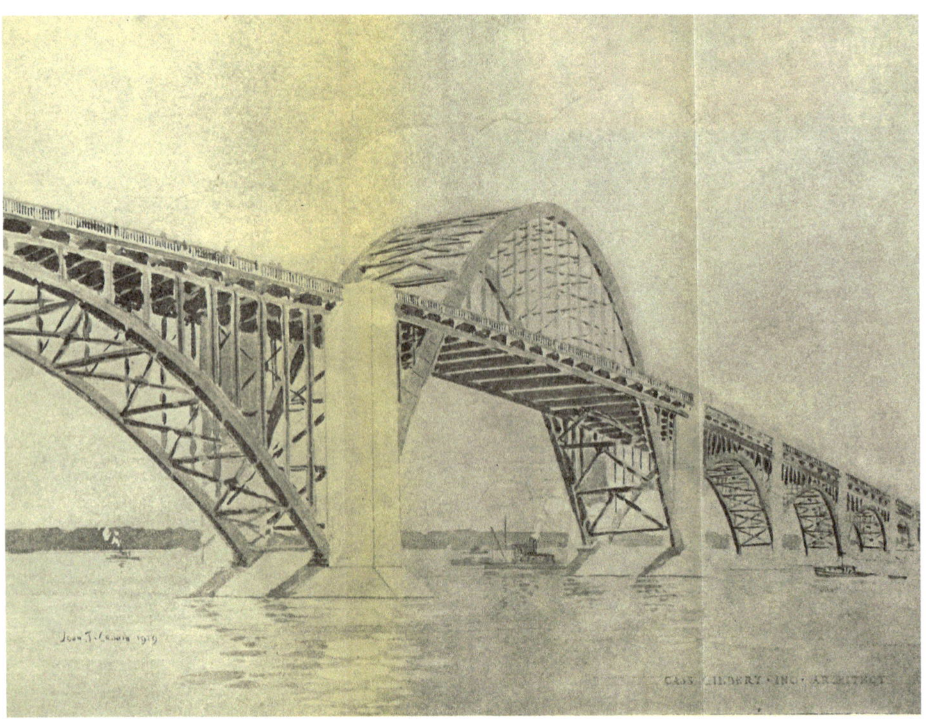

A Cass Gilbert firm architectural drawing of the proposed North Grand Island Bridge, 1929. (*H&H*)

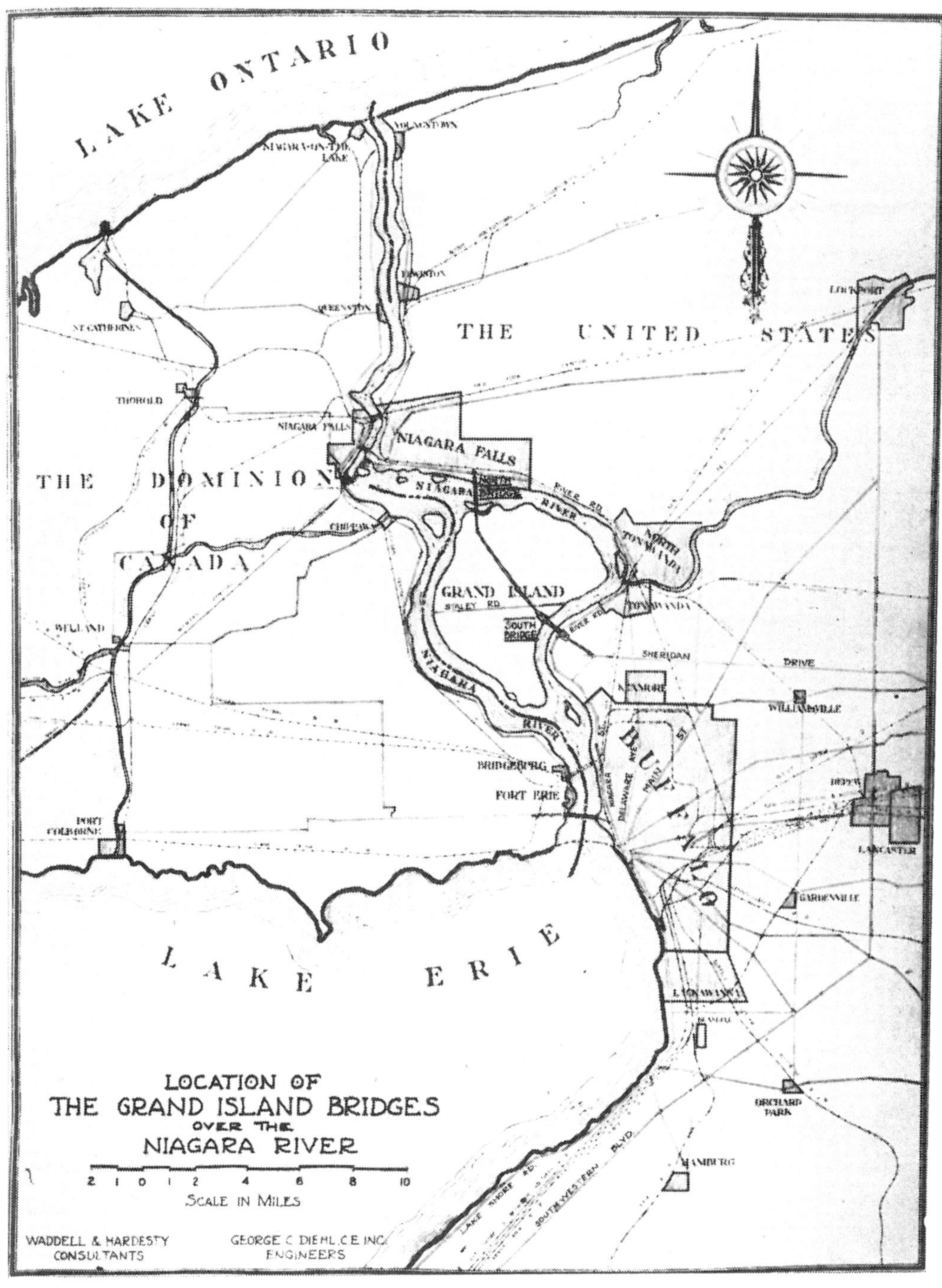

A Waddell & Hardesty map of the Niagara Falls area, showing the proposed location of the North and South Grand Island bridges, 1929. (*H&H*)

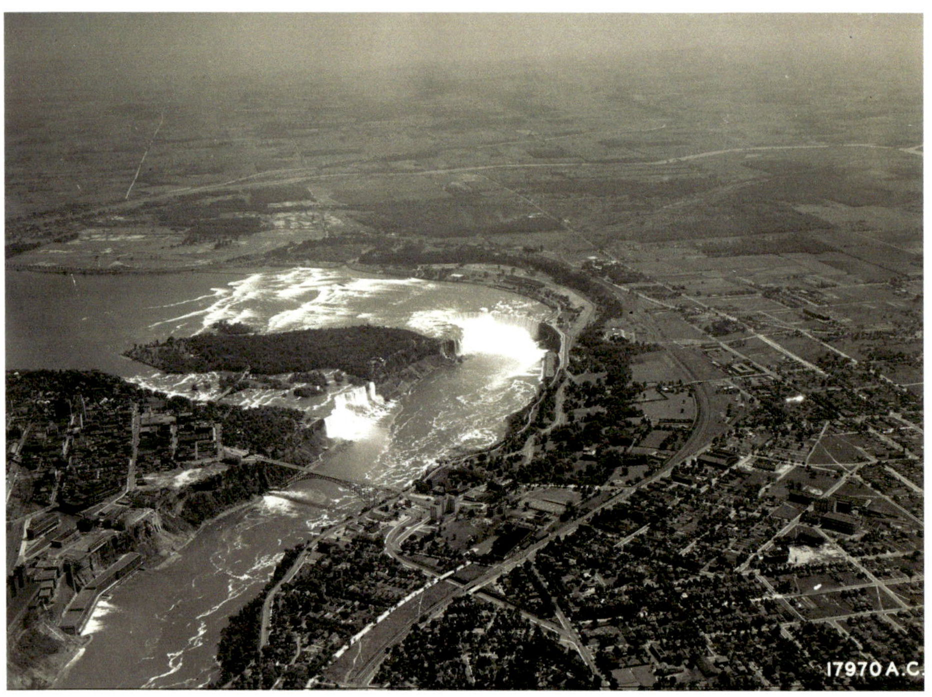

An aerial view of Niagara Falls taken by an Army photographer on October 15, 1931. (*National Archives*)

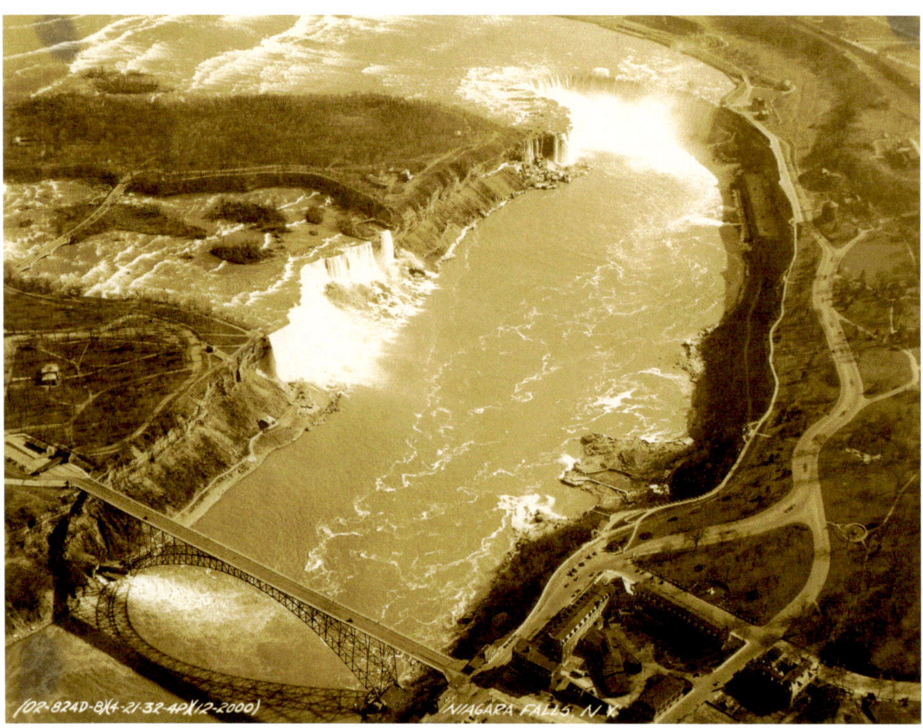

An aerial view of the Falls View Bridge in 1932. (*National Archives*)

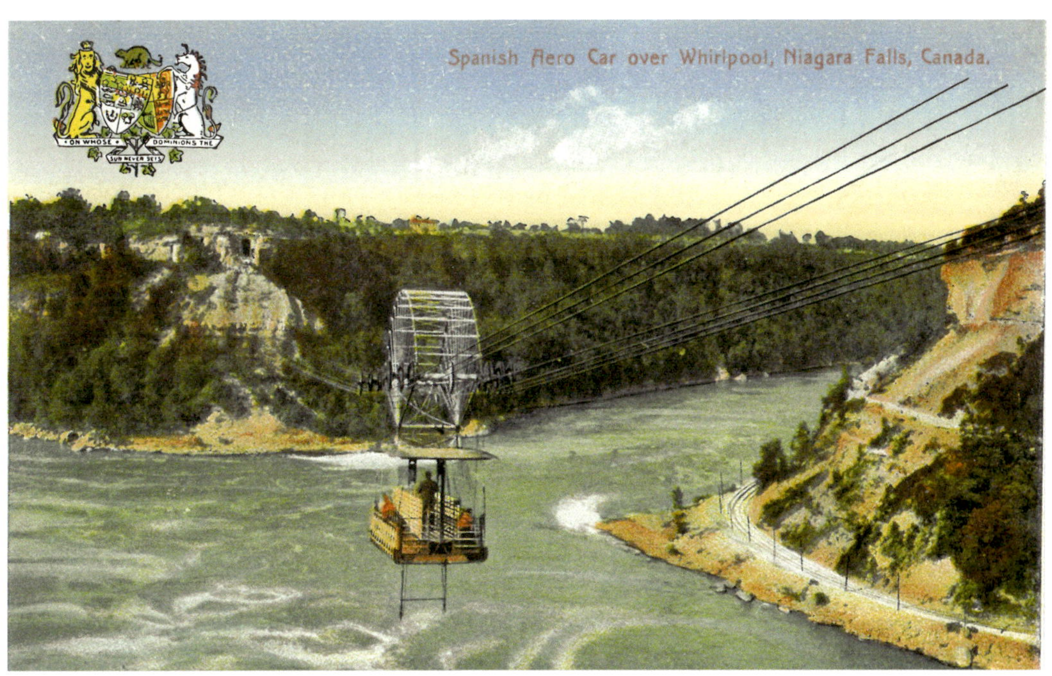

In 1913, a Spanish company obtained a fifty-year lease to build and operate a cable car ride over the rapids. This photo shows a *c.* 1930s view of a Spanish aero car over the Whirlpool Rapids.

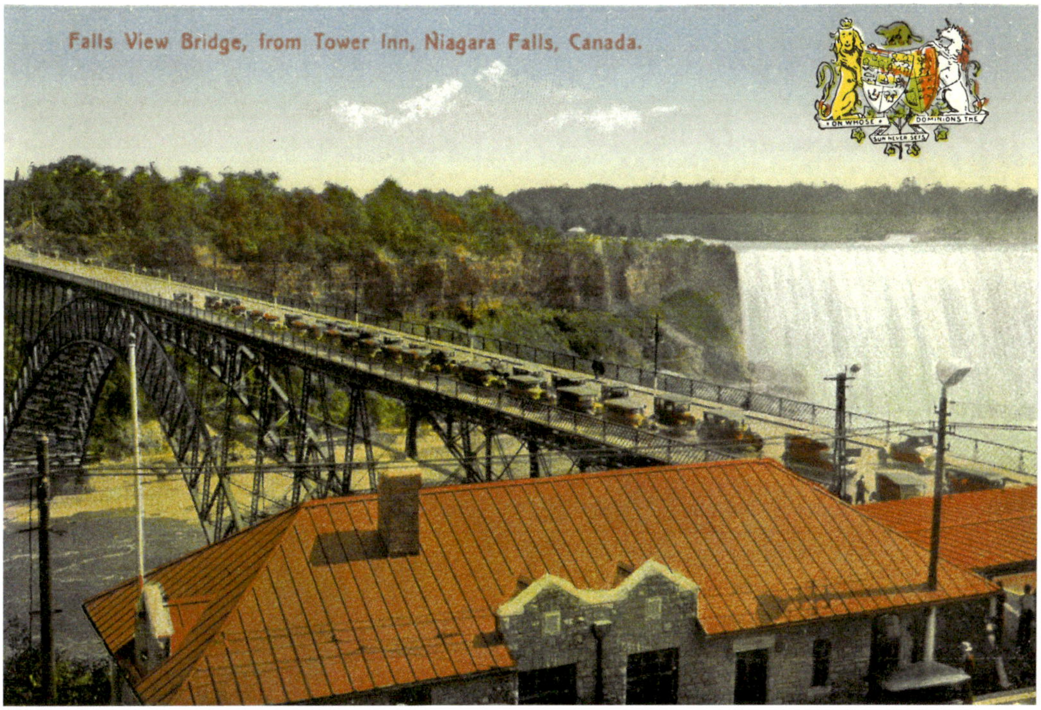

A *c.* 1930s view of the Falls View Bridge from the Tower Inn in Ontario.

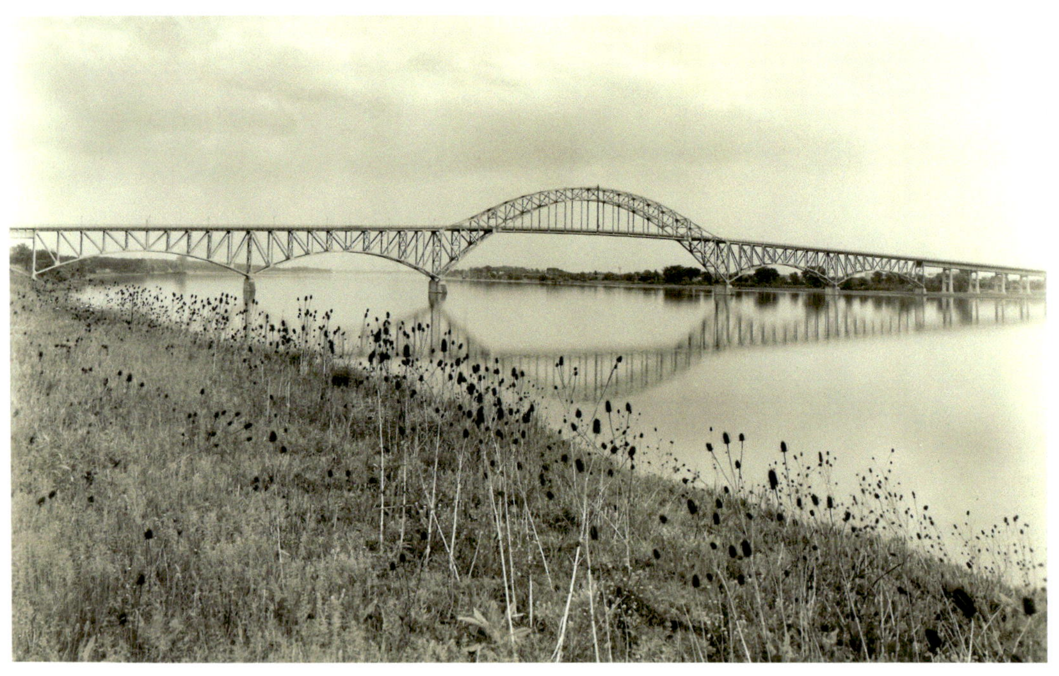

A peaceful view newly built of the South Grand Island Bridge over the Niagara River in 1935. (*H&H*)

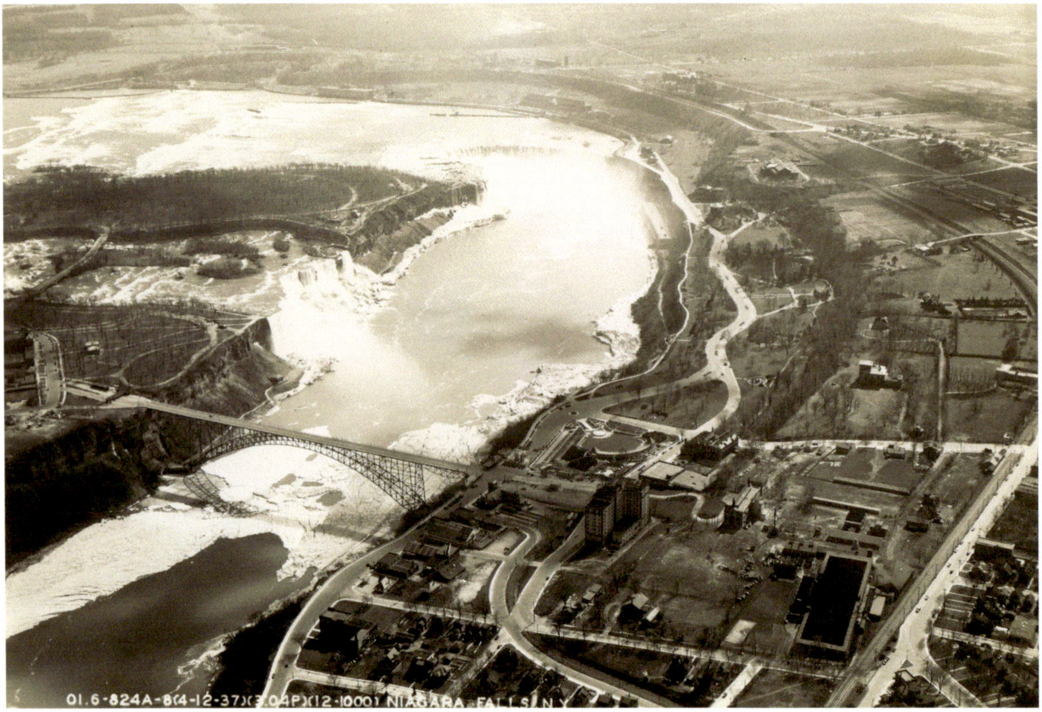

A 1937 aerial view of the precursor to the Rainbow Bridge, not long before it came down. The American side is to the left. (*National Archives*)

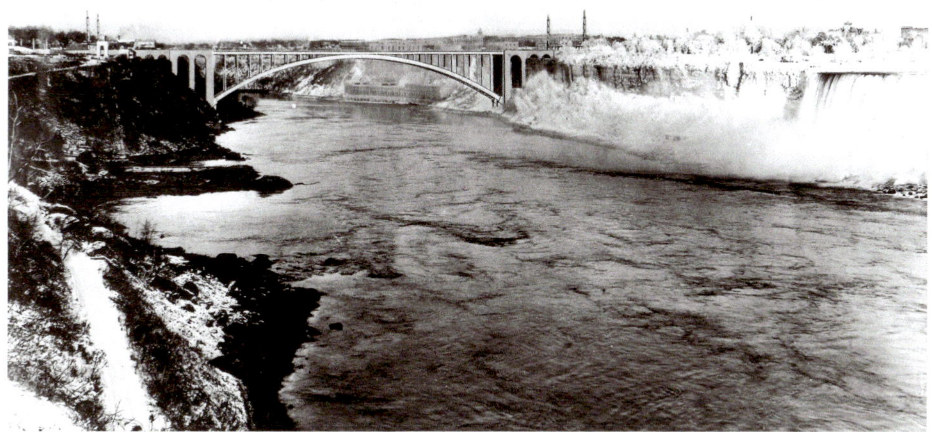

A rendering of the planned Rainbow Bridge, *c.* 1939, by the engineering firm of Waddell & Hardesty. The construction of a new bridge on the site was a priority after the old one had been destroyed by ice. (*H&H*)

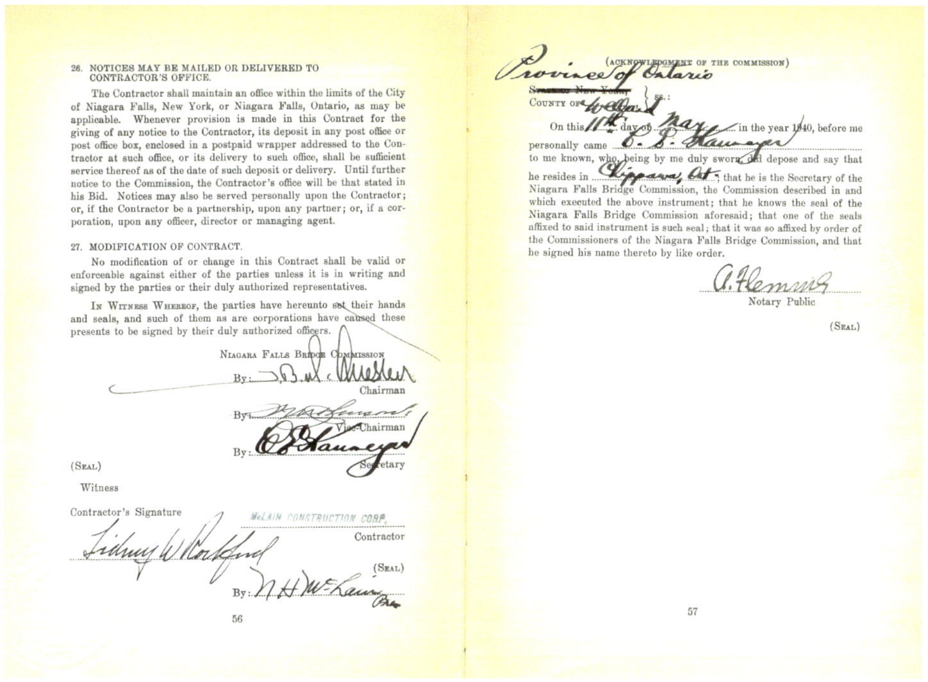

The signature pages from a 1940 contract for the Rainbow Bridge, signed by one of the contractors—McLain Construction Corporation—as well as three members of the Niagara Falls Bridge Commission: the chairman, vice chairman, and secretary. (*H&H*)

NIAGARA FALLS BRIDGE COMMISSION

RAINBOW BRIDGE OVER NIAGARA RIVER

BETWEEN NIAGARA FALLS, NEW YORK,
AND NIAGARA FALLS, ONTARIO

INFORMATION FOR BIDDERS, BID, CONTRACT,
BOND, AND SPECIFICATIONS

CONTRACTS NO. 3A, 3B, 4, AND 5

SUPERSTRUCTURE

WADDELL & HARDESTY, NEW YORK, NEW YORK
EDWARD P. LUPFER CORPORATION, BUFFALO, NEW YORK
CONSULTING ENGINEERS

HAGEY & GRAY ENGINEERING COMPANY
FORT ERIE, ONTARIO
ASSOCIATED CANADIAN ENGINEERS

FEBRUARY, 1940

The front cover of the bid documents book for Contracts 3A, 3B, 4, and 5 for the superstructure of the Rainbow Bridge, February 1940. (*H&H*)

NIAGARA FALLS BRIDGE COMMISSION

RAINBOW BRIDGE OVER NIAGARA RIVER

BETWEEN NIAGARA FALLS, NEW YORK,
AND NIAGARA FALLS, ONTARIO

INFORMATION FOR BIDDERS, BID, CONTRACT,
BOND, AND SPECIFICATIONS

CONTRACT NO. 6

LIGHTING SYSTEM ON AMERICAN SIDE

WADDELL & HARDESTY, NEW YORK, NEW YORK
EDWARD P. LUPFER CORPORATION, BUFFALO, NEW YORK
CONSULTING ENGINEERS

HAGEY & GRAY ENGINEERING COMPANY
FORT ERIE, ONTARIO
ASSOCIATED CANADIAN ENGINEERS

JULY, 1941

Shortridge Hardesty's (of the lead design firm Waddell & Hardesty) copy of the bid documents for Contract No. 6 of the Rainbow Bridge, issued in July 1941, for the lighting system on the American side. (*H&H*)

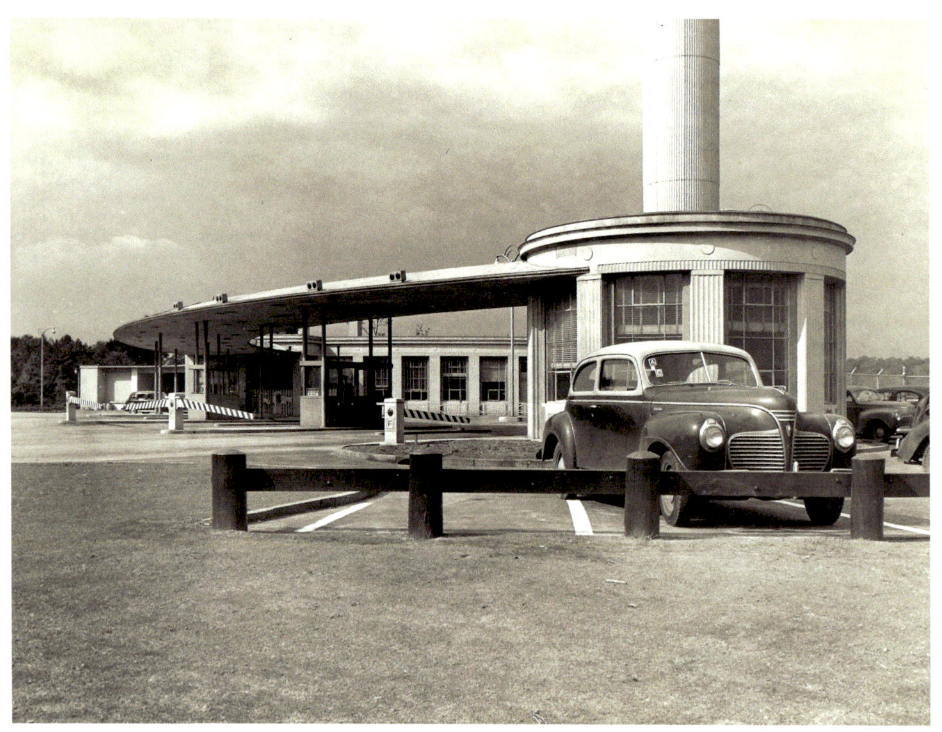

The newly opened Rainbow Bridge toll plaza, 1941. (*H&H*)

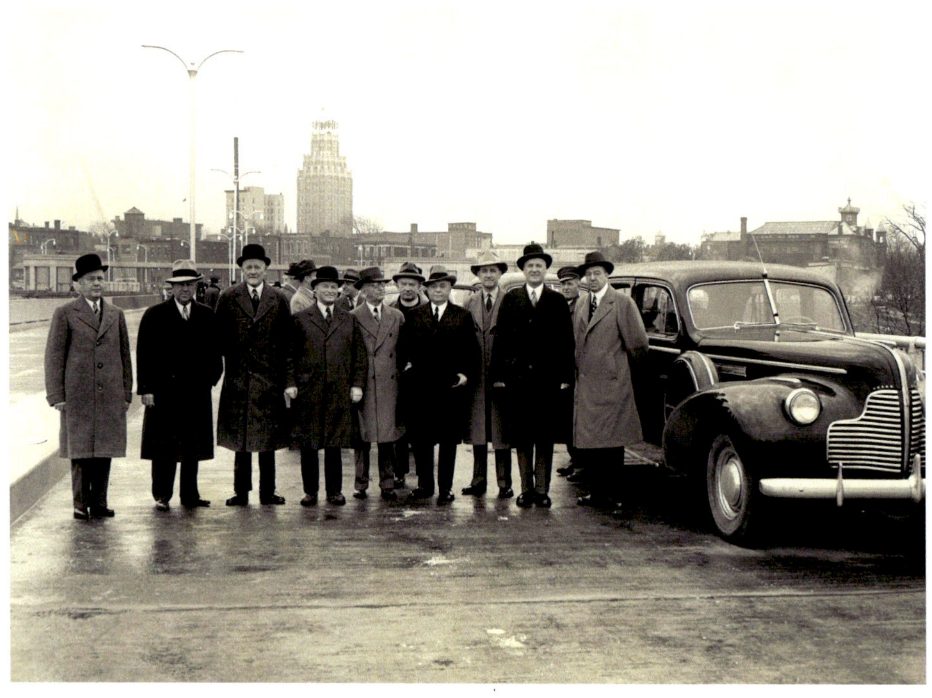

A group of officials pose during the opening of the Rainbow Bridge in 1941. (*H&H*)

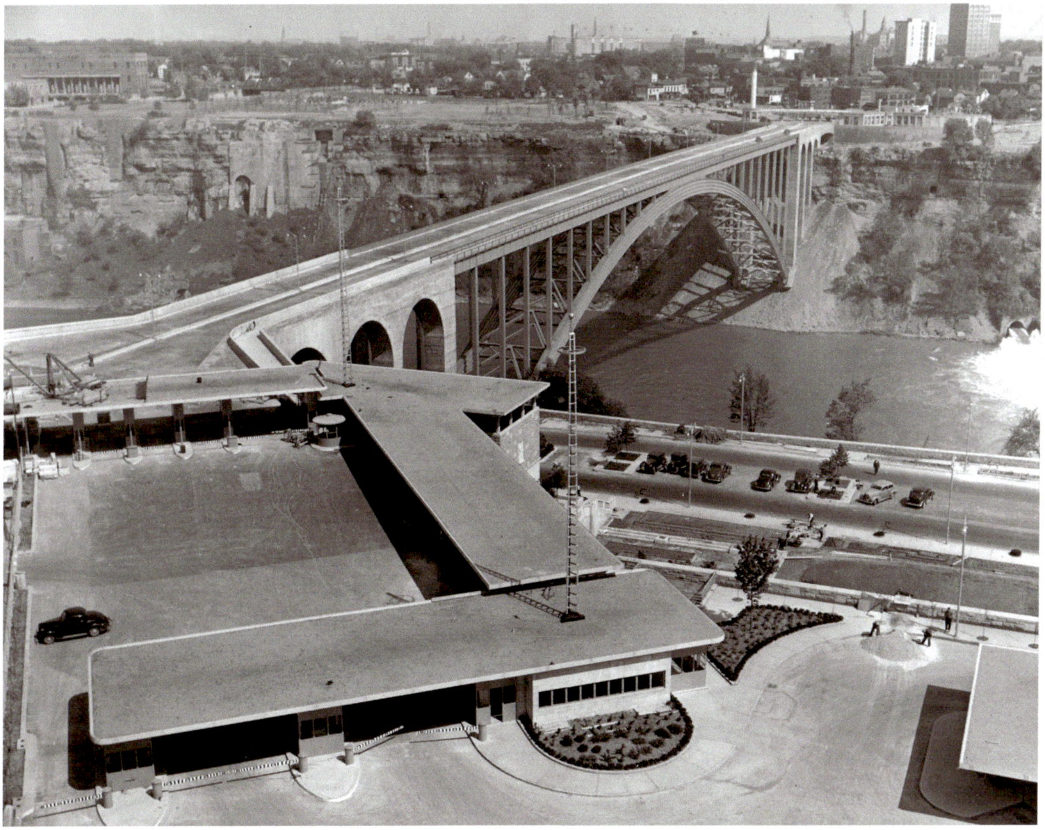

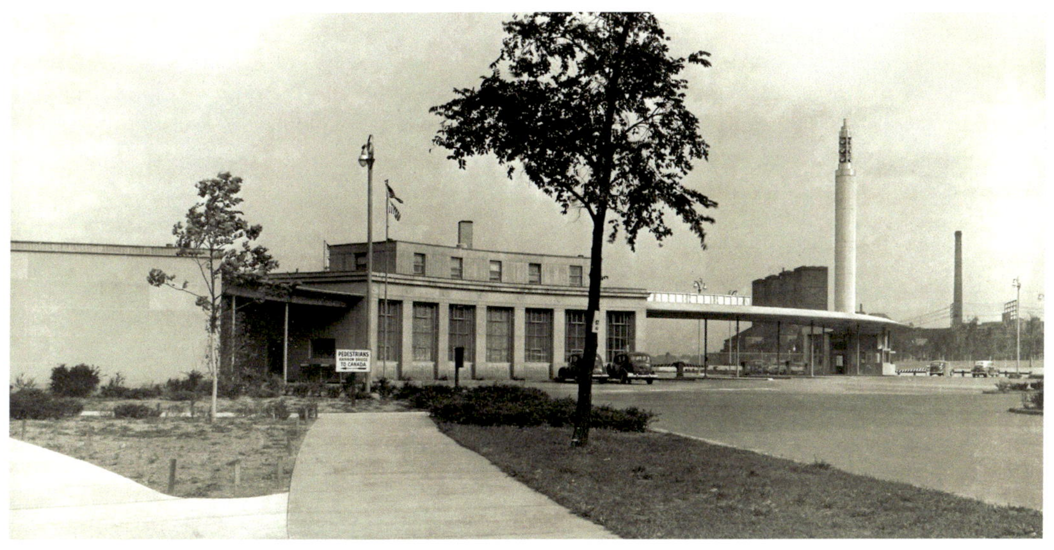

An aerial view of the newly completed Rainbow Bridge from the Canadian side, showing the concrete approach arches over River Road, 1941. (*H&H*)

An American side Rainbow Bridge administration building, 1941. (*H&H*)

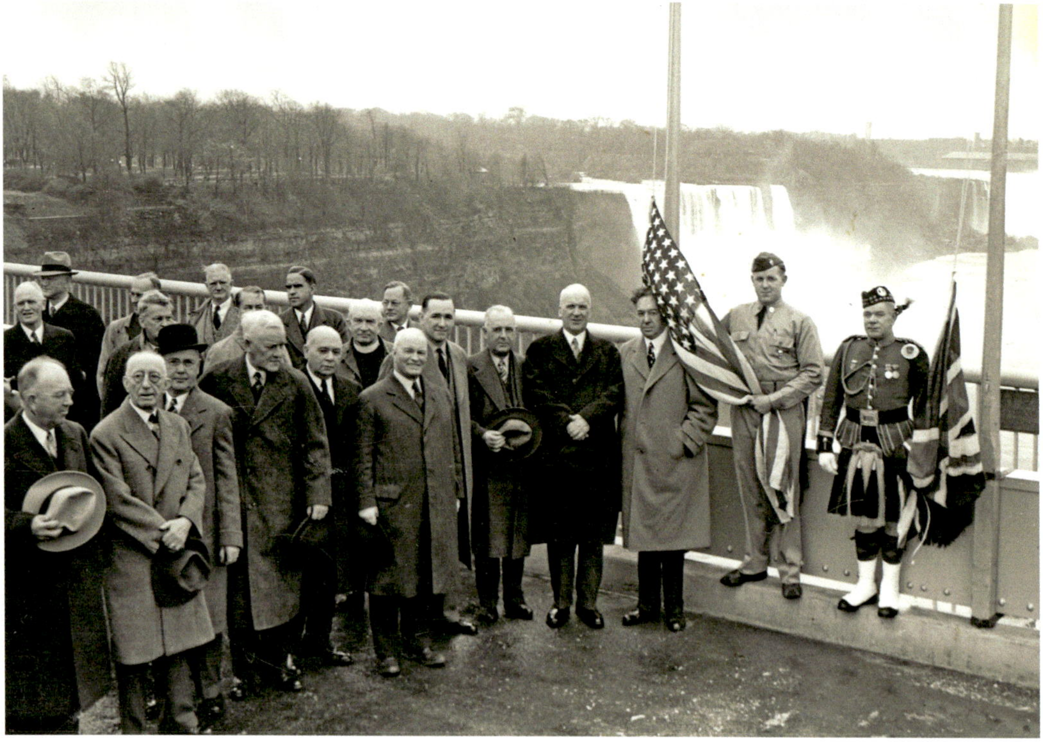

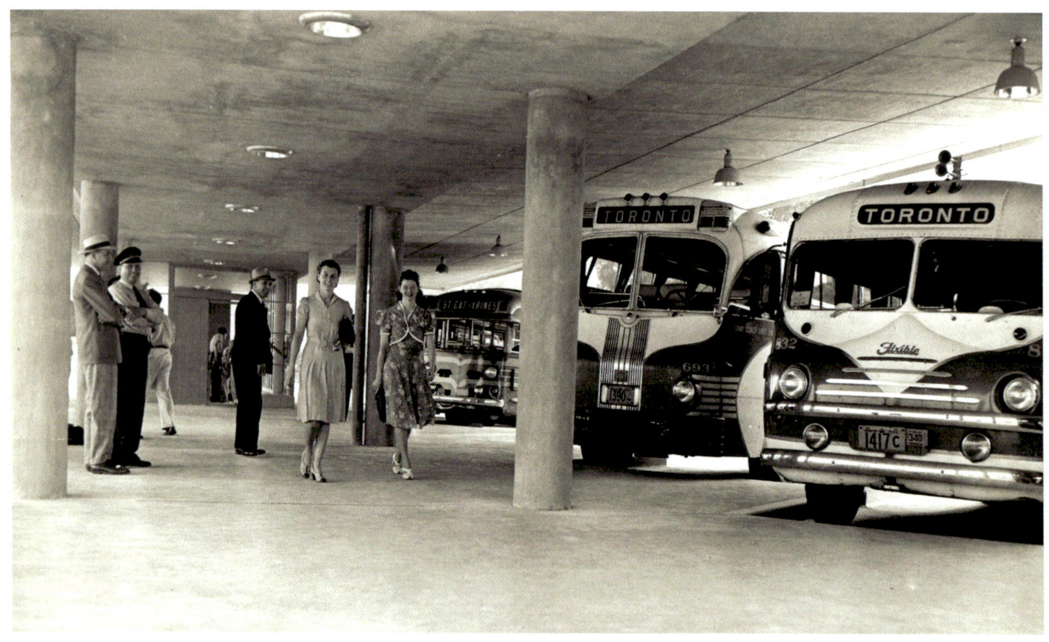

Many Canadian and American officials attended the flag-raising ceremony at the dedication of the Rainbow Bridge, November 3, 1941. (*H&H*)

Tourist buses parked near the Rainbow Bridge, 1941. (*H&H*)

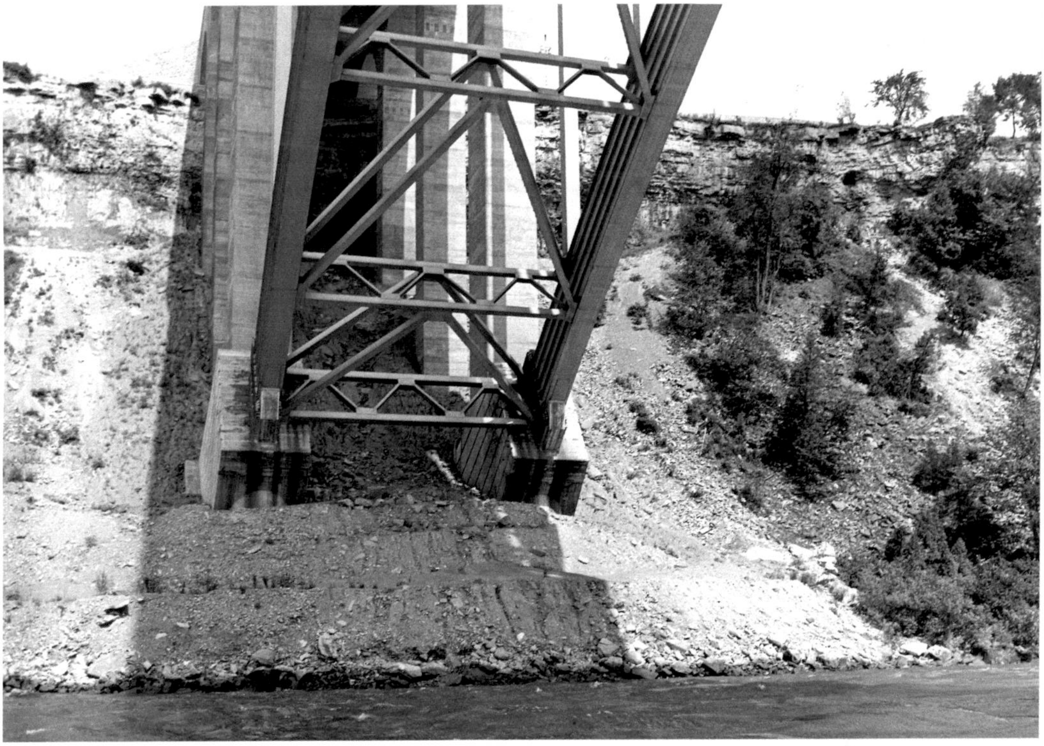

Close-up of one of the abutments of the Rainbow Bridge, 1941. (*H&H*)

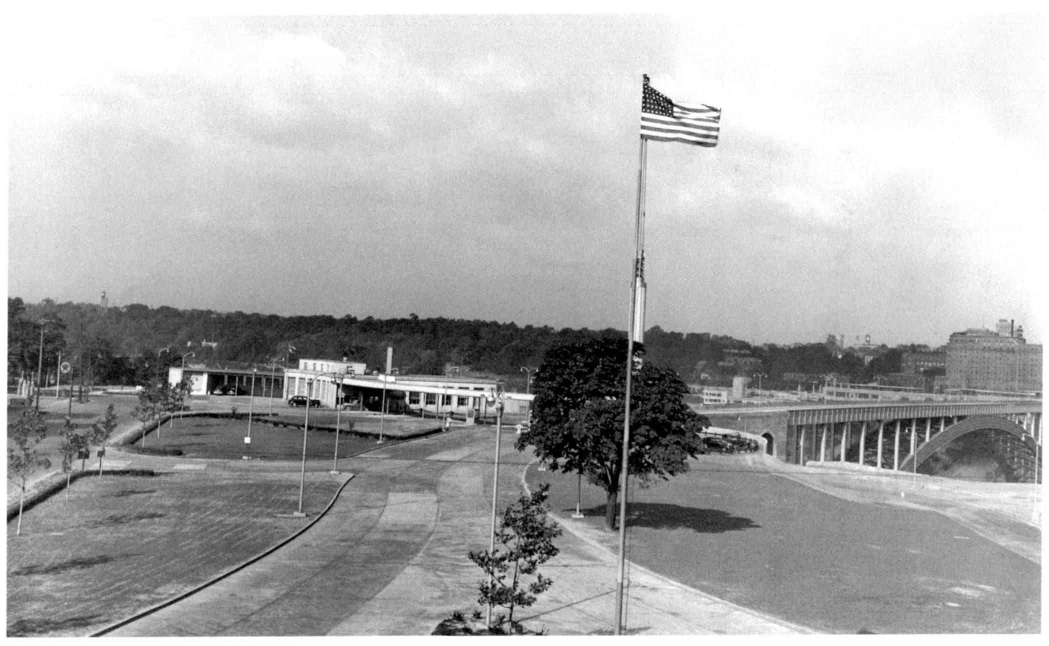

A 1941 view of the American approach to the newly constructed Rainbow Bridge. (*H&H*)

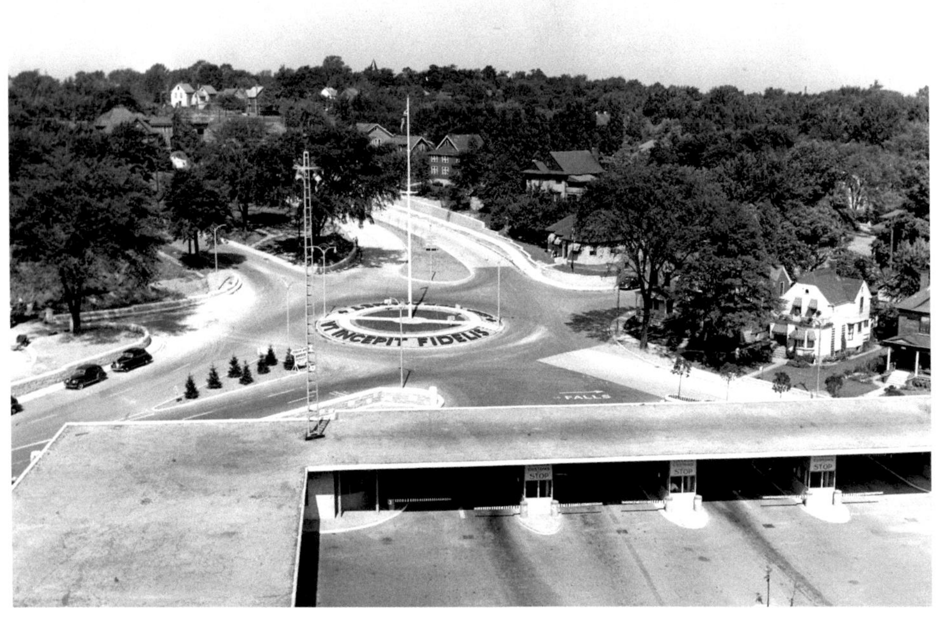

Above and below: The Canadian toll plaza of the newly constructed Rainbow Bridge, 1941. (*H&H*)

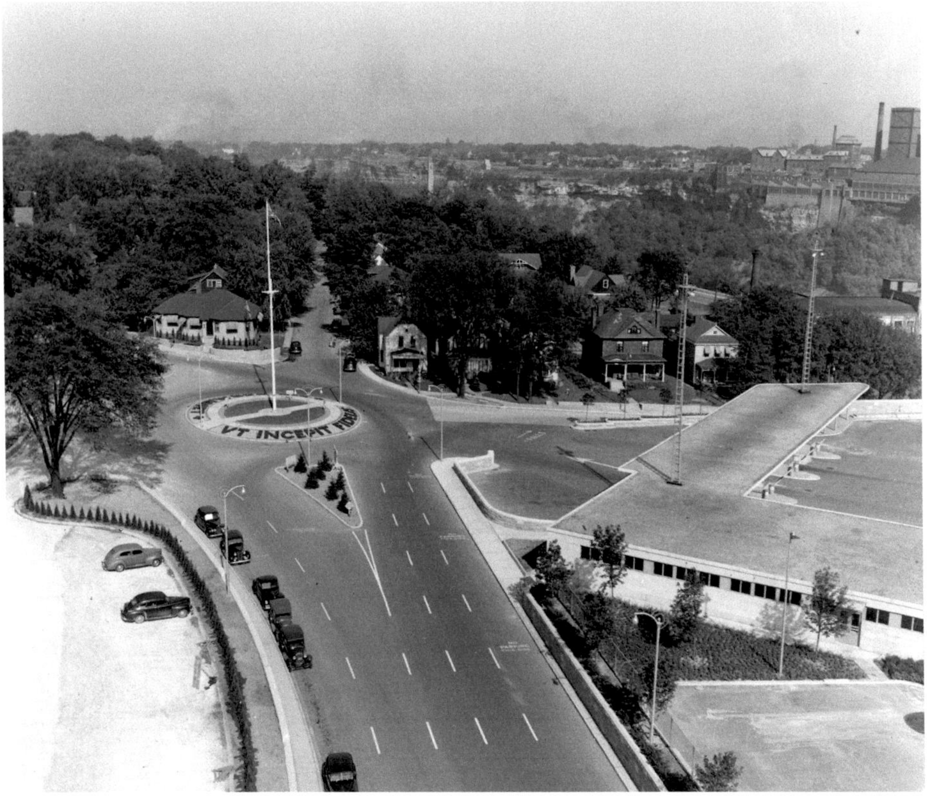

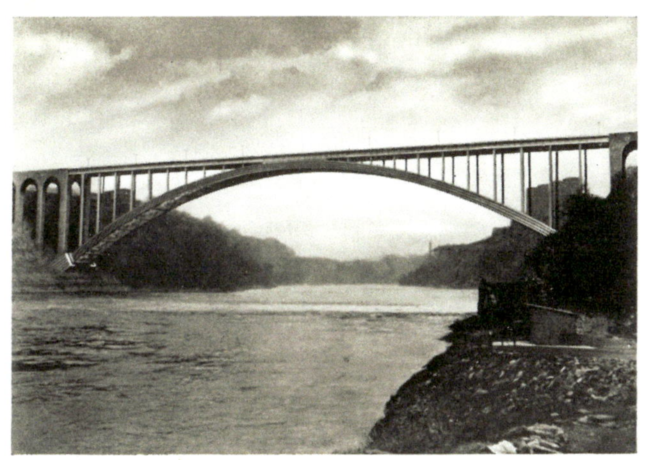

**FIRST PLACE—CLASS A—
1941 AWARD**

RAINBOW BRIDGE—Over Niagara River, between Niagara Falls, New York and Niagara Falls, Ontario, Canada; Total Cost, $4,000,000; Engineers: Waddell & Hardesty; Edward P. Lupfer Corporation; Architect: Aymar Embury II; Fabricators and Erectors: Bethlehem Steel Company; Owner: Niagara Falls Bridge Commission; Date Completed and Opened to Traffic, November 1, 1941; Span length: Main span—hingeless arch—960 feet.

The Rainbow Bridge, constructed at a cost of $4 million, won a First Place Class A award from the American Institute of Steel Construction (AISC) in 1941. Shown here is a page from the prize booklet. (*H&H*)

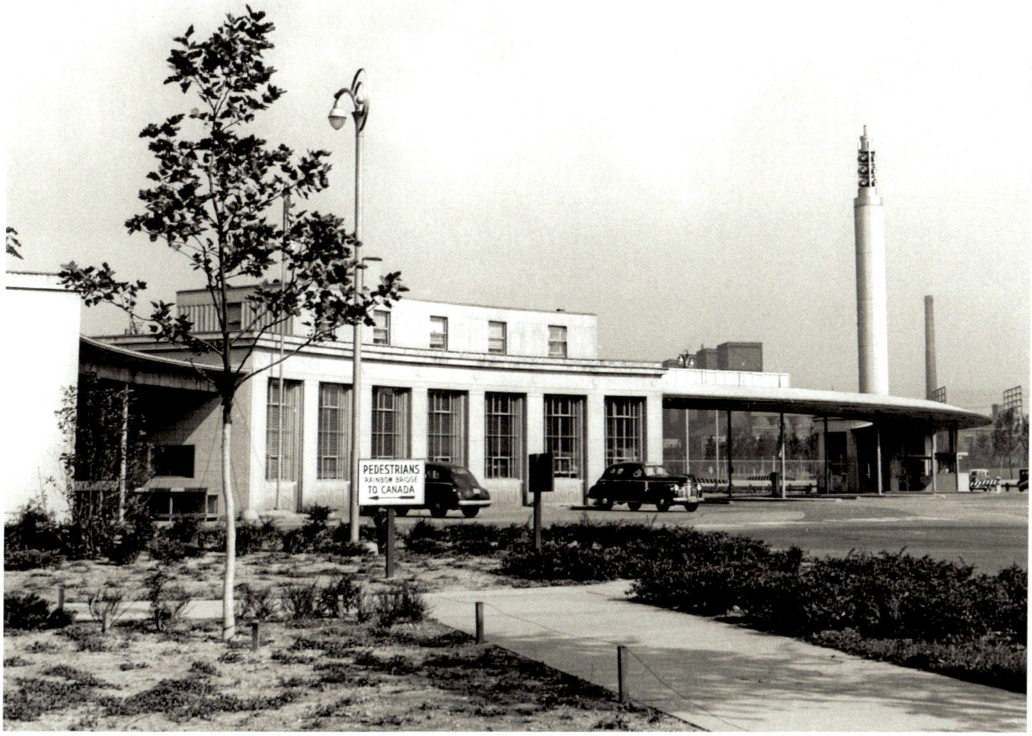

Close-up of an administration building at the Rainbow Bridge, 1941. (*H&H*)

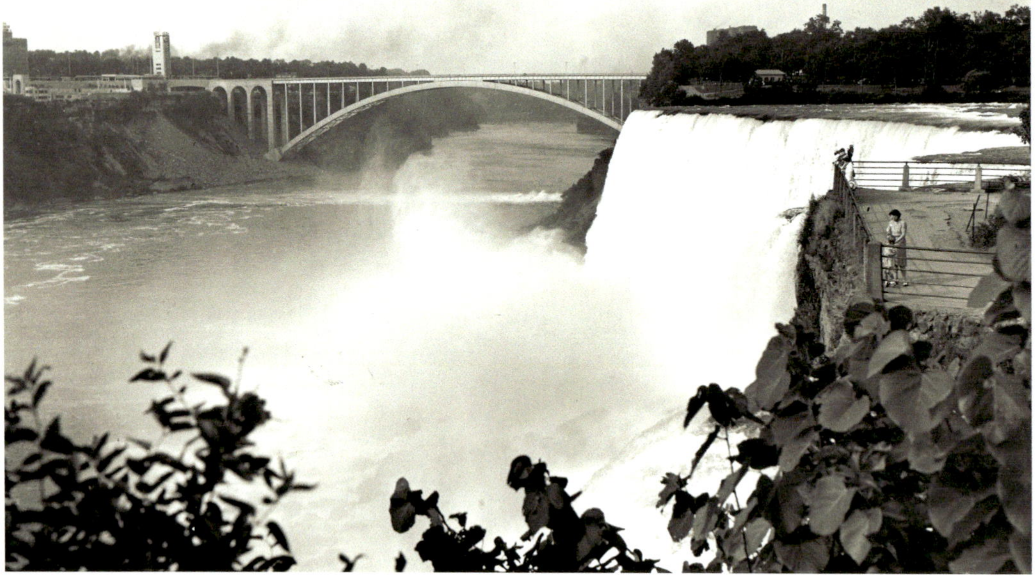

Above and below: Two *c.* 1940s views of the Rainbow Bridge as seen through the mists of the falls. (*H&H*)

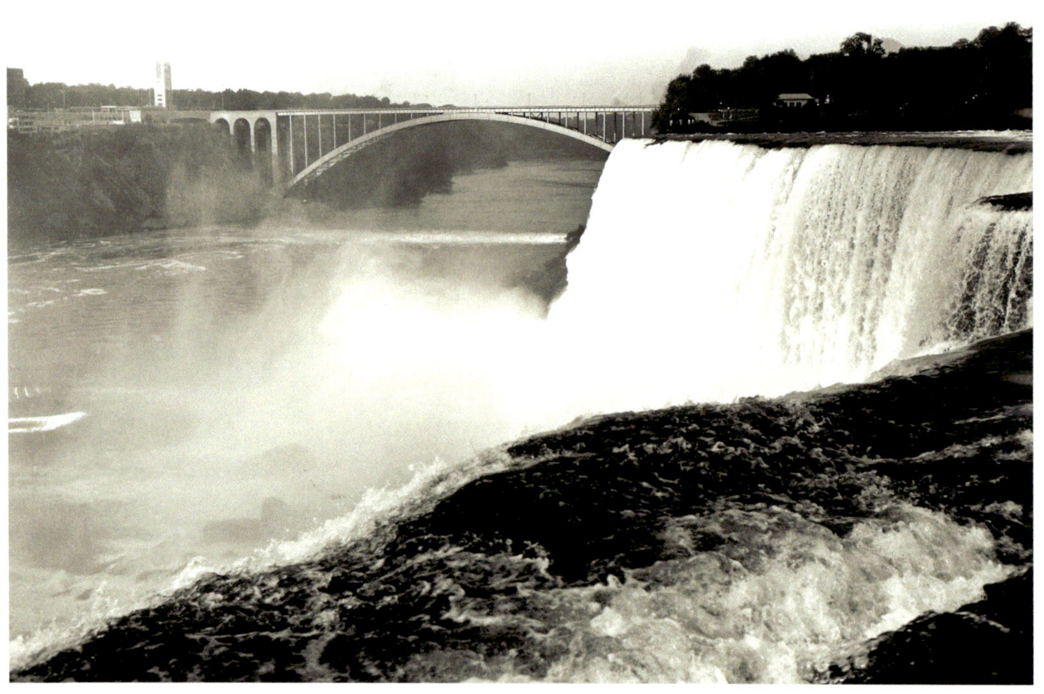

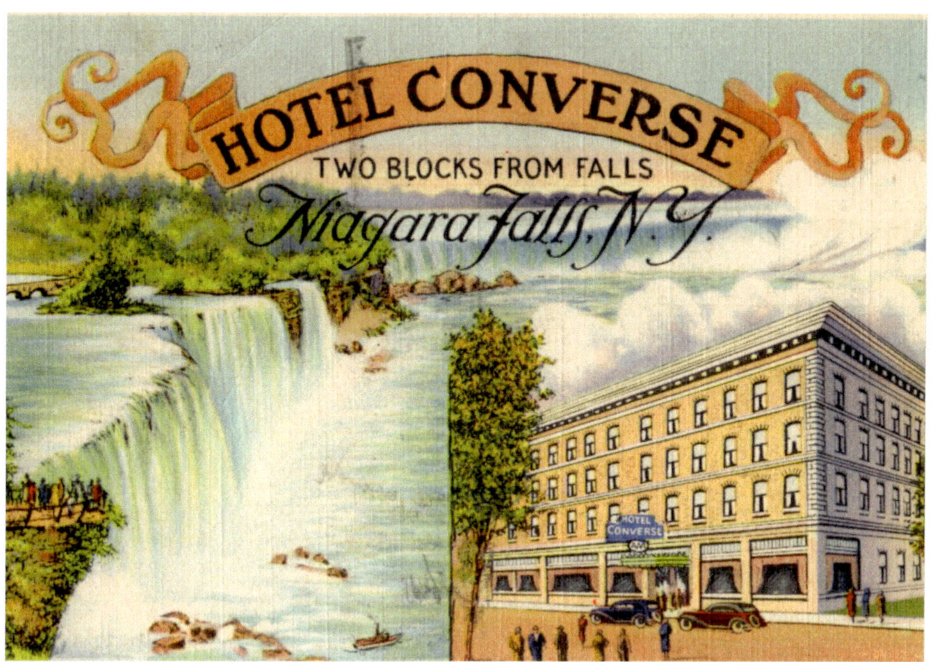

A *c.* 1940s color postcard of the Hotel Converse overlooking the falls. The Goat Island Bridge can be seen at the far left.

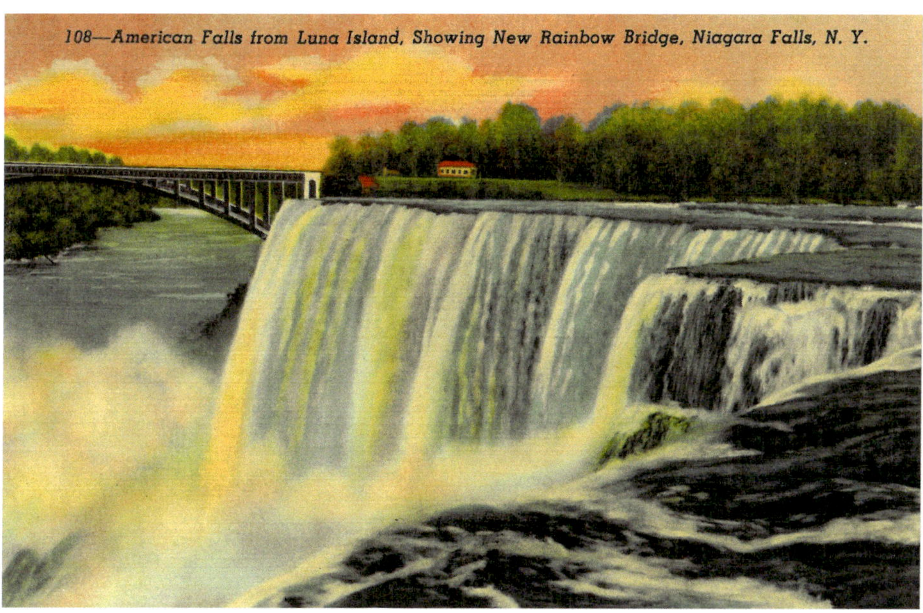

A *c.* 1940s postcard of Niagara Falls with the Rainbow Bridge in the background. The caption on the reverse explains: "The contour of Niagara Falls is constantly changing from the effects of frost, erosion and undermining of the rocks by falling water. Because of these workings of nature, the American Falls have been changed in the past few years from almost a straight front to the ragged crest line shown in this picture."

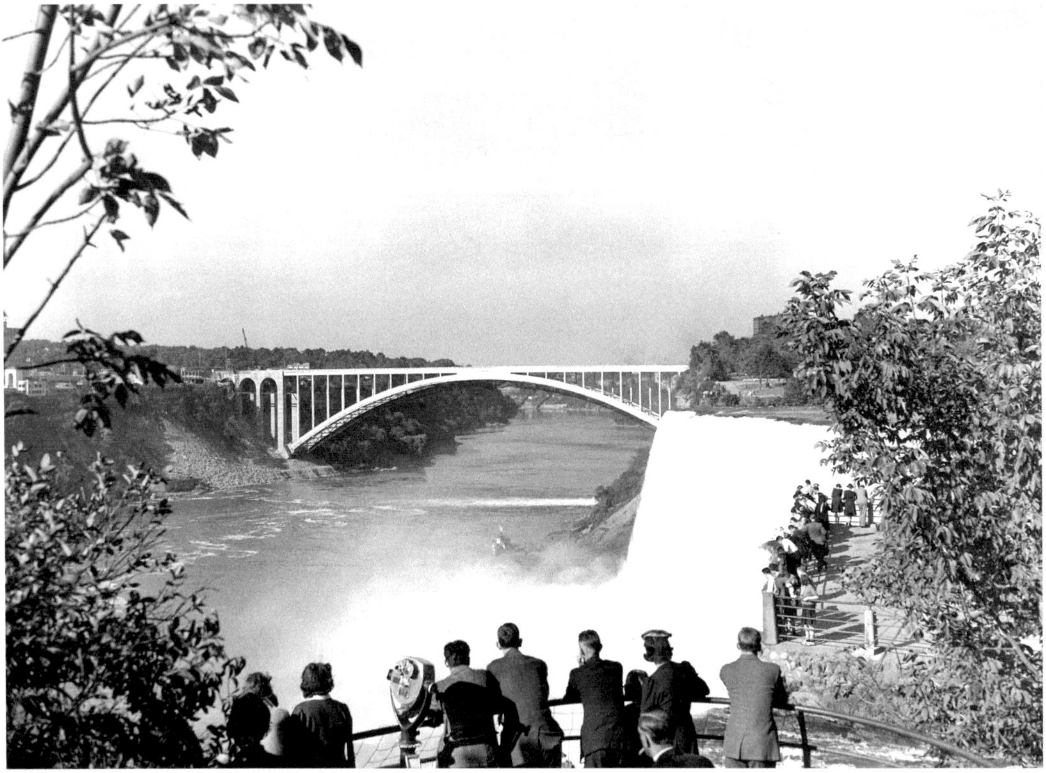

A group of people take in the falls and the Rainbow Bridge from an observation platform, *c.* 1940s. (*H&H*)

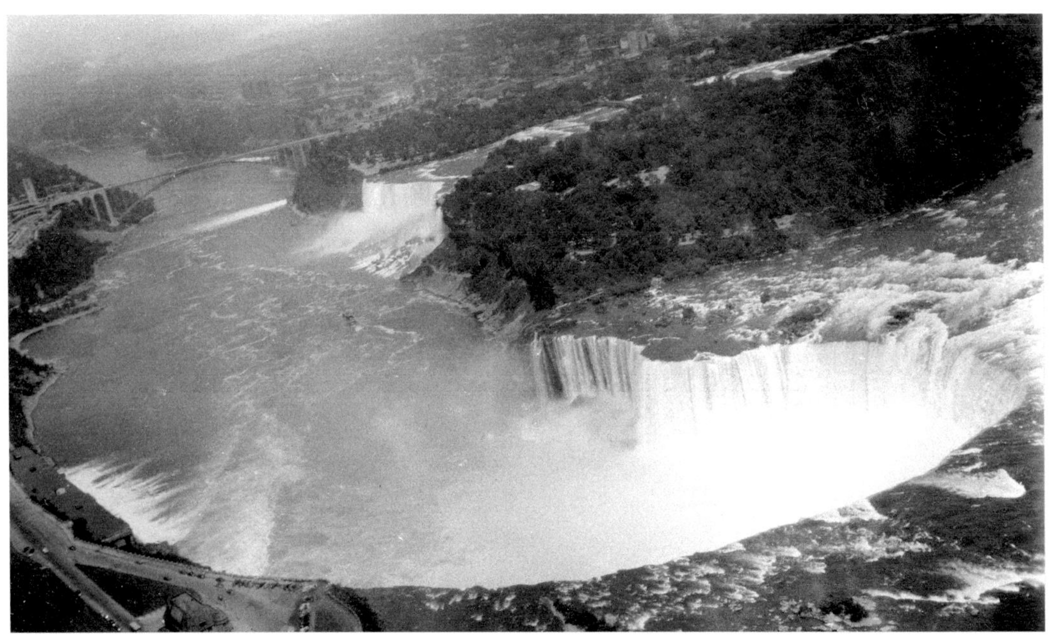

An aerial view of Niagara Falls with the Rainbow Bridge in the background, July 1949. (*H&H*)

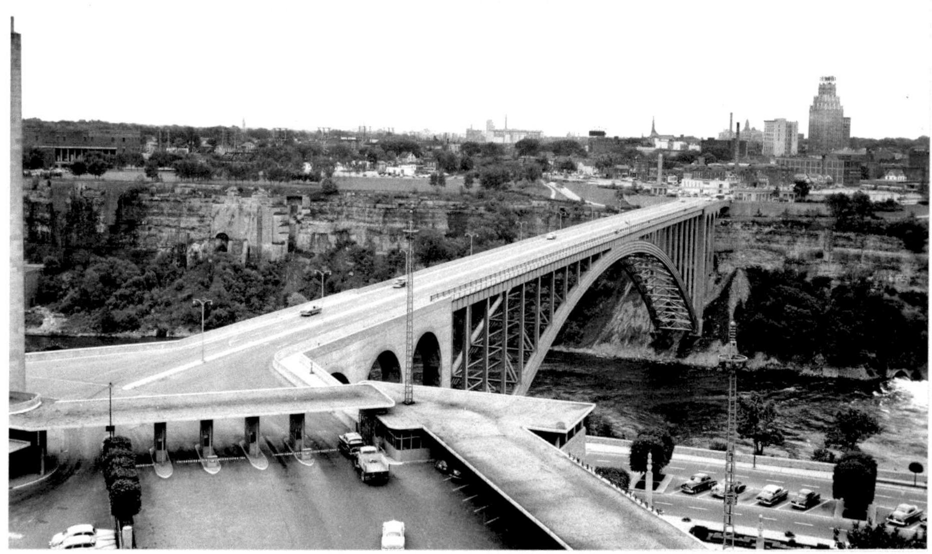

The Rainbow Bridge as seen from the New York side in June 1952, looking toward Ontario. (*National Archives*)

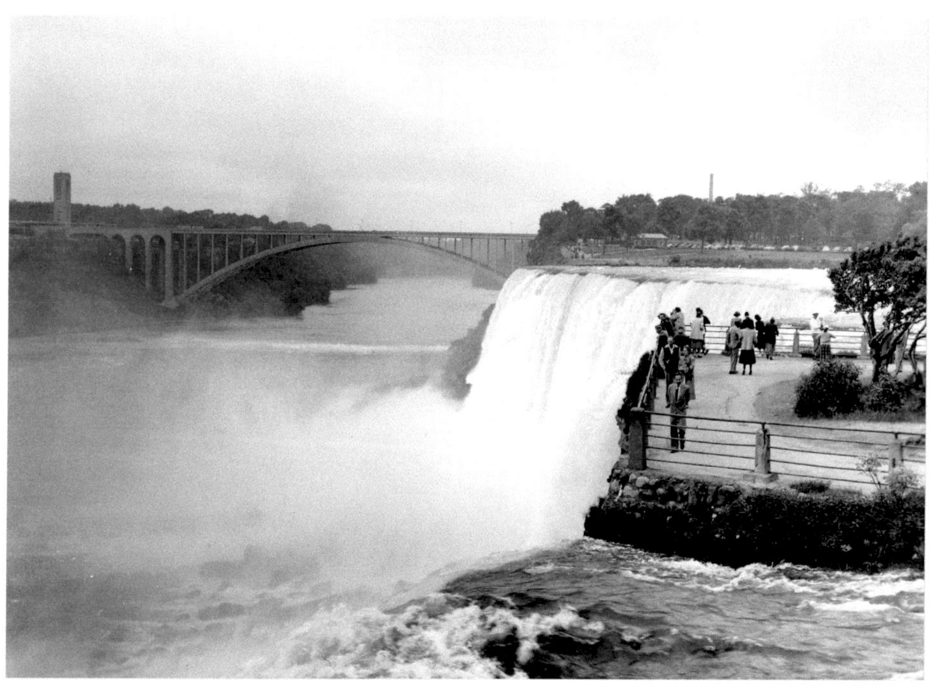

View of Niagara Falls and the Rainbow Bridge from Luna Island, June 1952. [*Photo by T. W. Kines*] (*National Archives*)

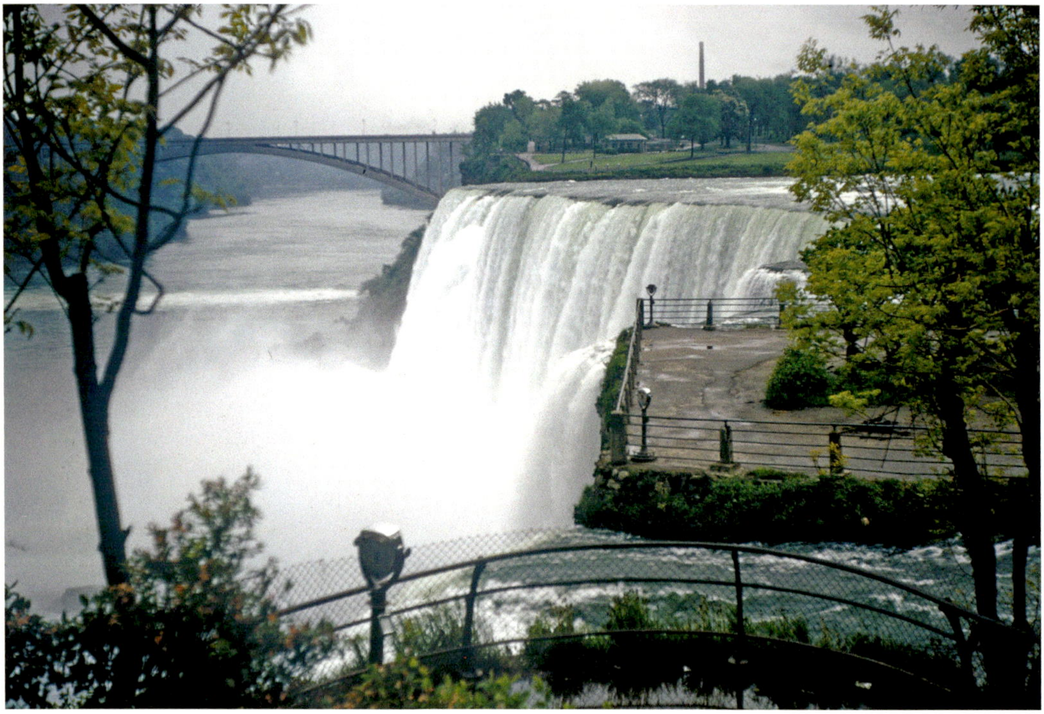

The American Falls and Rainbow Bridge seen from an observation platform, May 31, 1953. (*US Geologic Survey*)

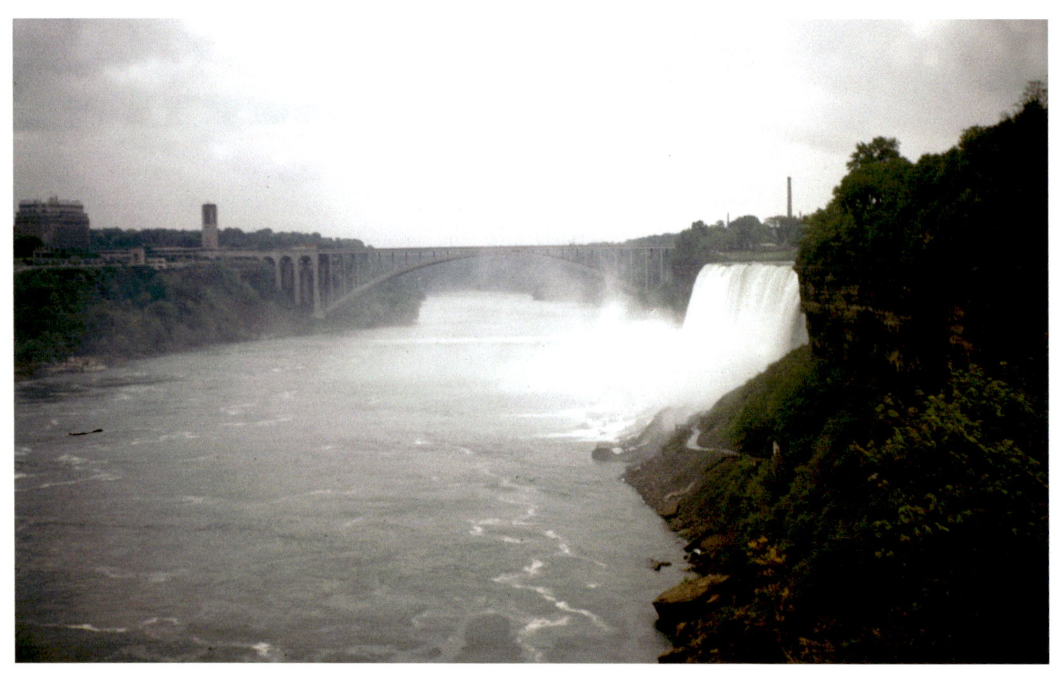

The American Falls as seen from Three Sisters, May 31, 1953. (*US Geologic Survey*)

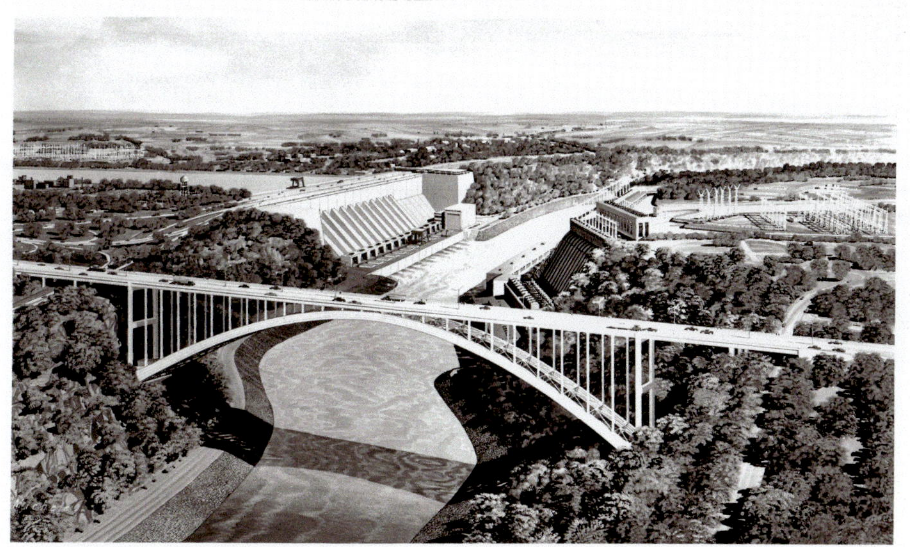

Above and below: A pair of *c.* 1960 renderings by New York City engineering firm Hardesty & Hanover of the proposed Lewiston-Queenston Bridge. (*H&H*)

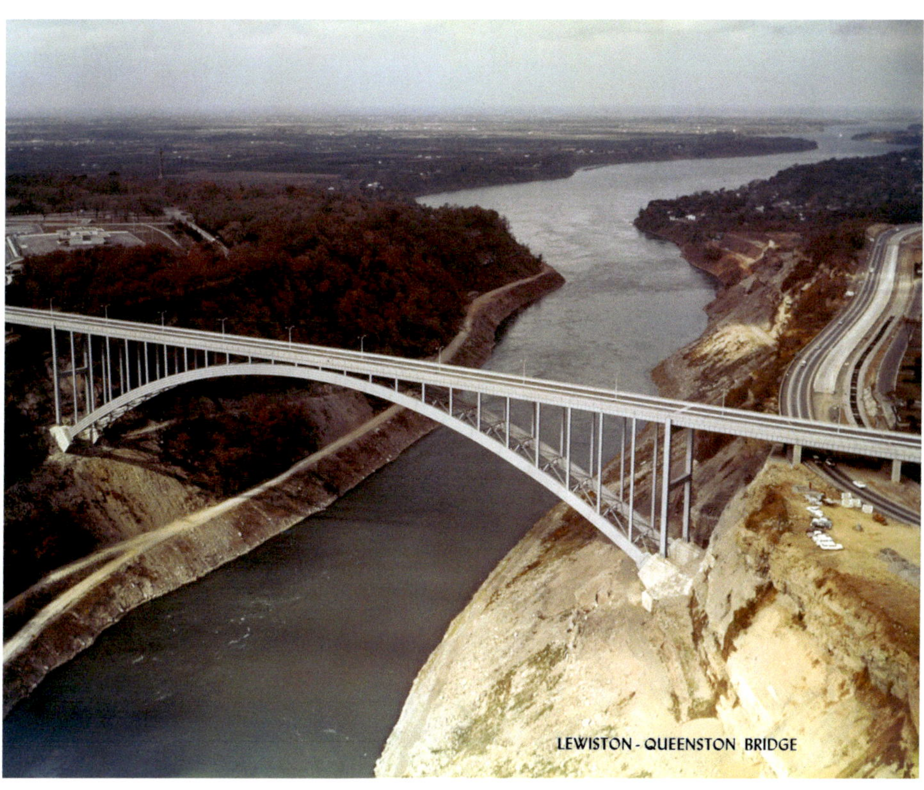

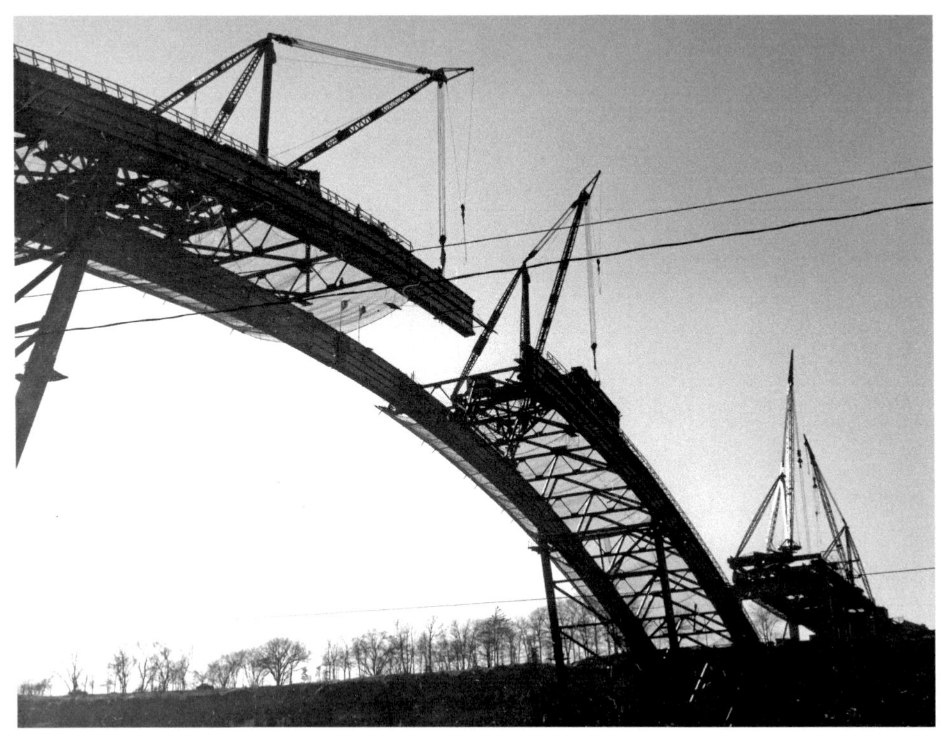

Above and below: Cranes on site at the Lewiston-Queenston Bridge as it is under construction, 1962. (*H&H*)

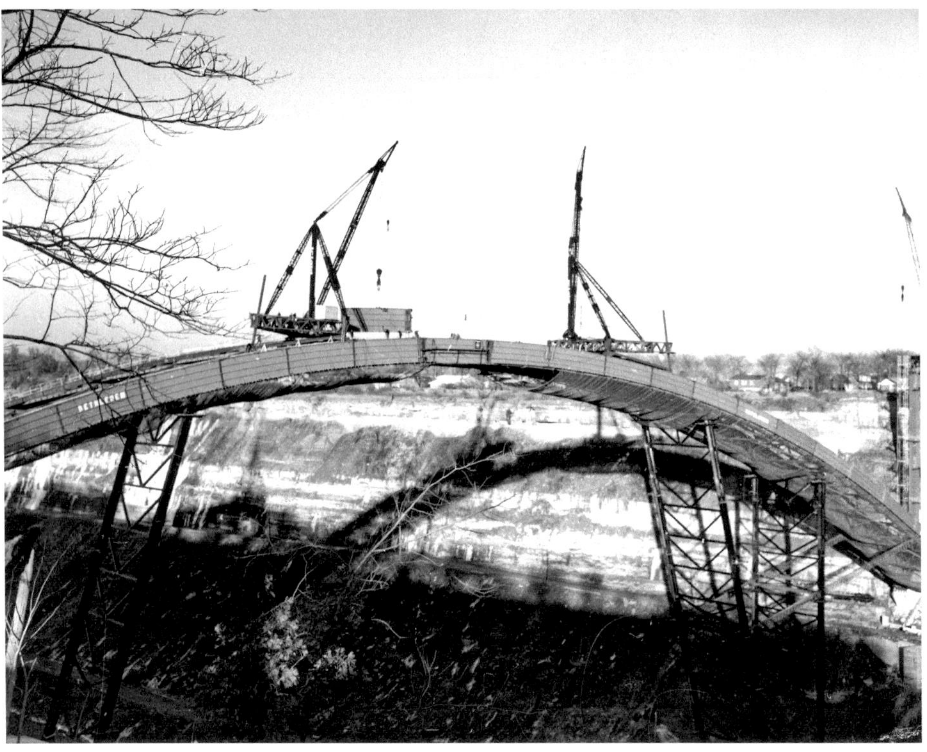

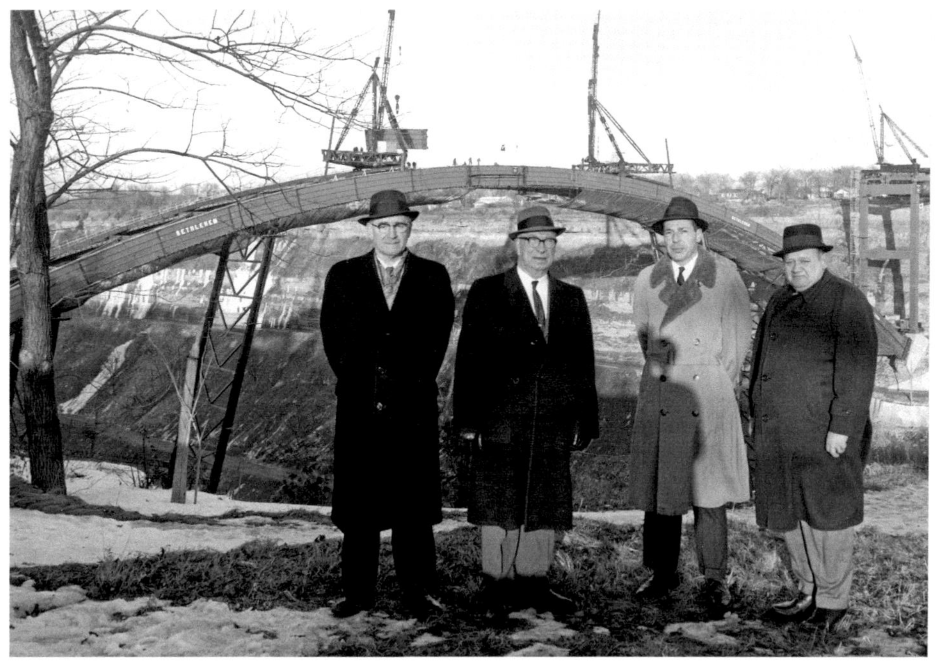

This photograph by Frank Seed shows bridge agency officials C. Ellison Kaumeyer, Hon. Charles Daley, John J. Bingenheimer, and James J. Upton near the site of the Lewiston-Queenston Bridge, January 1962. (*H&H*)

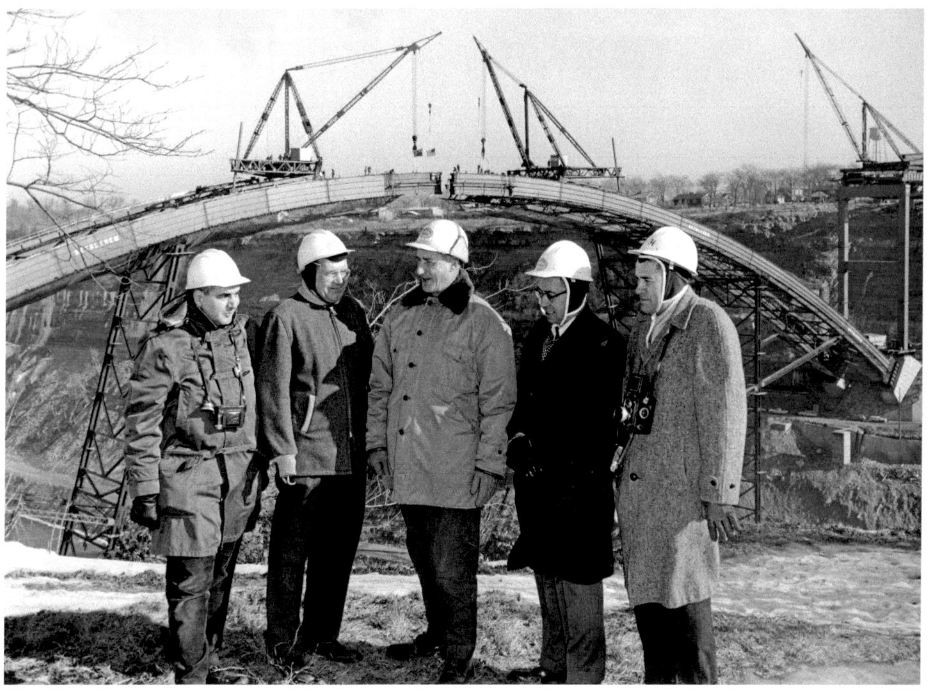

Engineers from the Lewiston-Queenston Bridge's design firm, Hardesty & Hanover, stand in front of the bridge as it is under construction in 1962. (*H&H*)

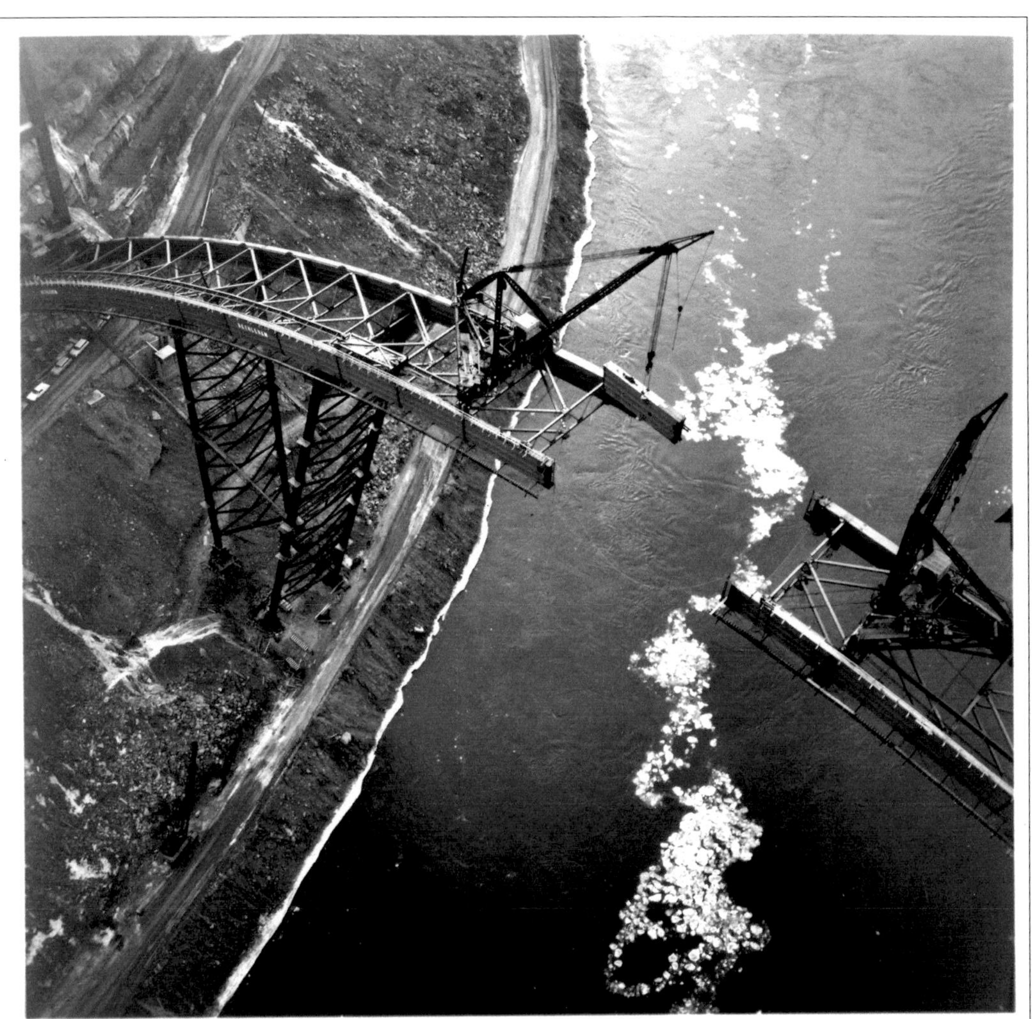

The Lewiston Queenston Bridge under construction as it appeared January 24, 1962, just as the arch was about to be completed. (*H&H*)

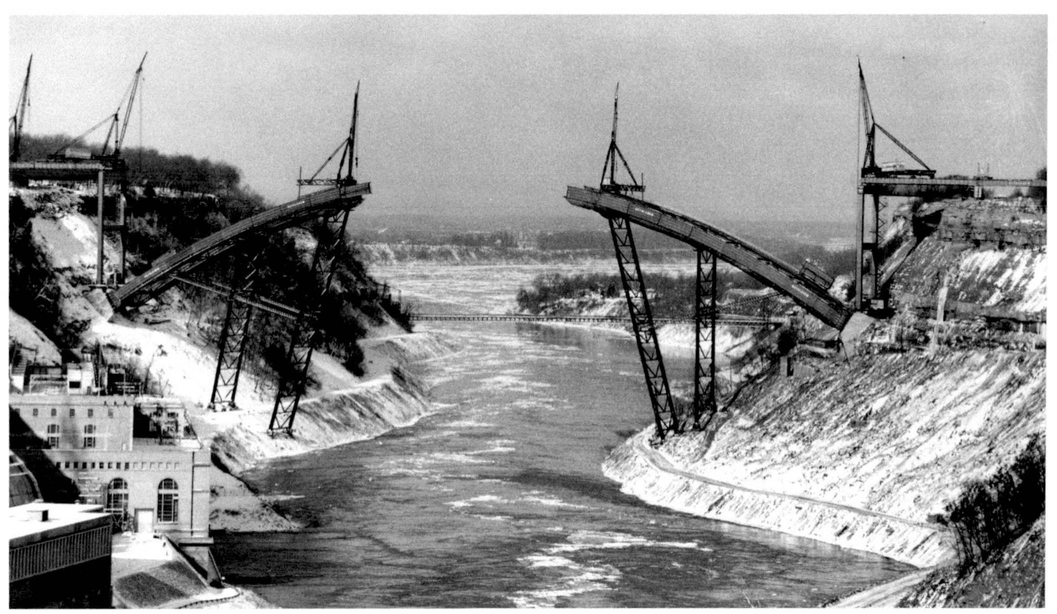

These four images taken in 1962 show the arch of the Lewiston-Queenston Bridge at various stages of completion. Note the temporary supports. (*H&H*)

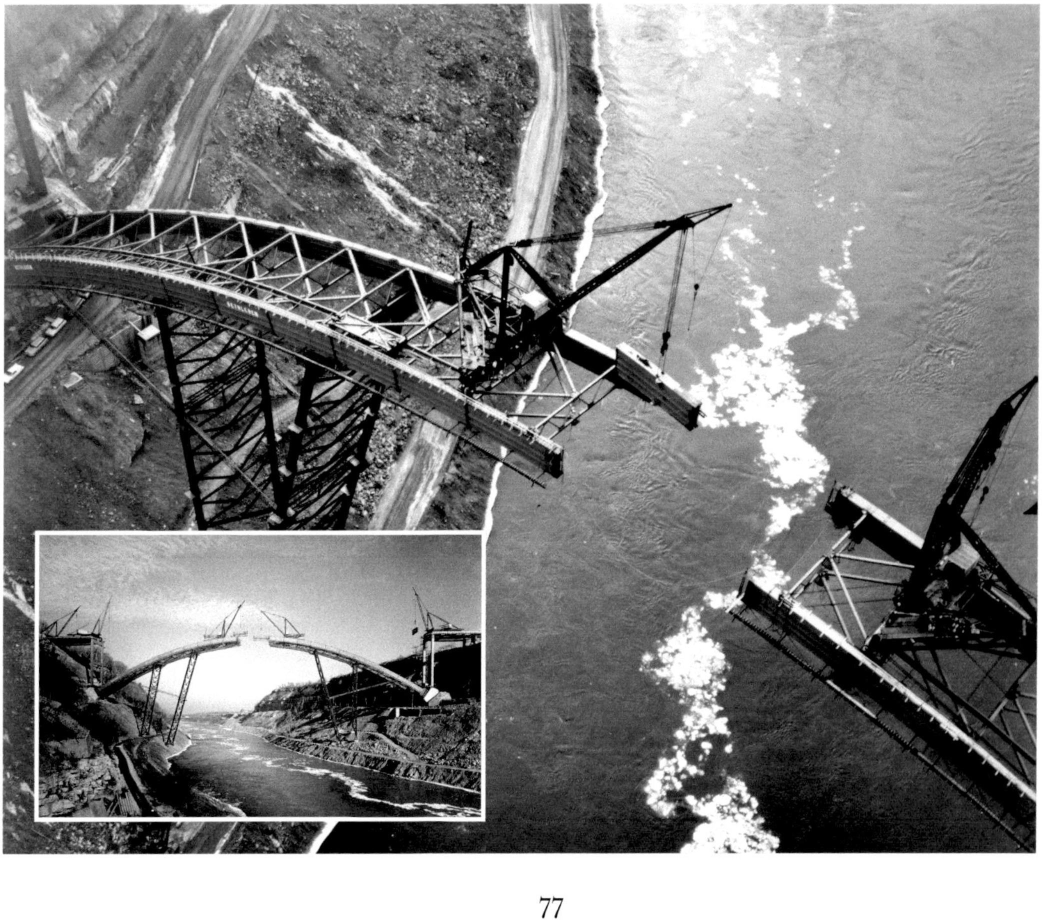

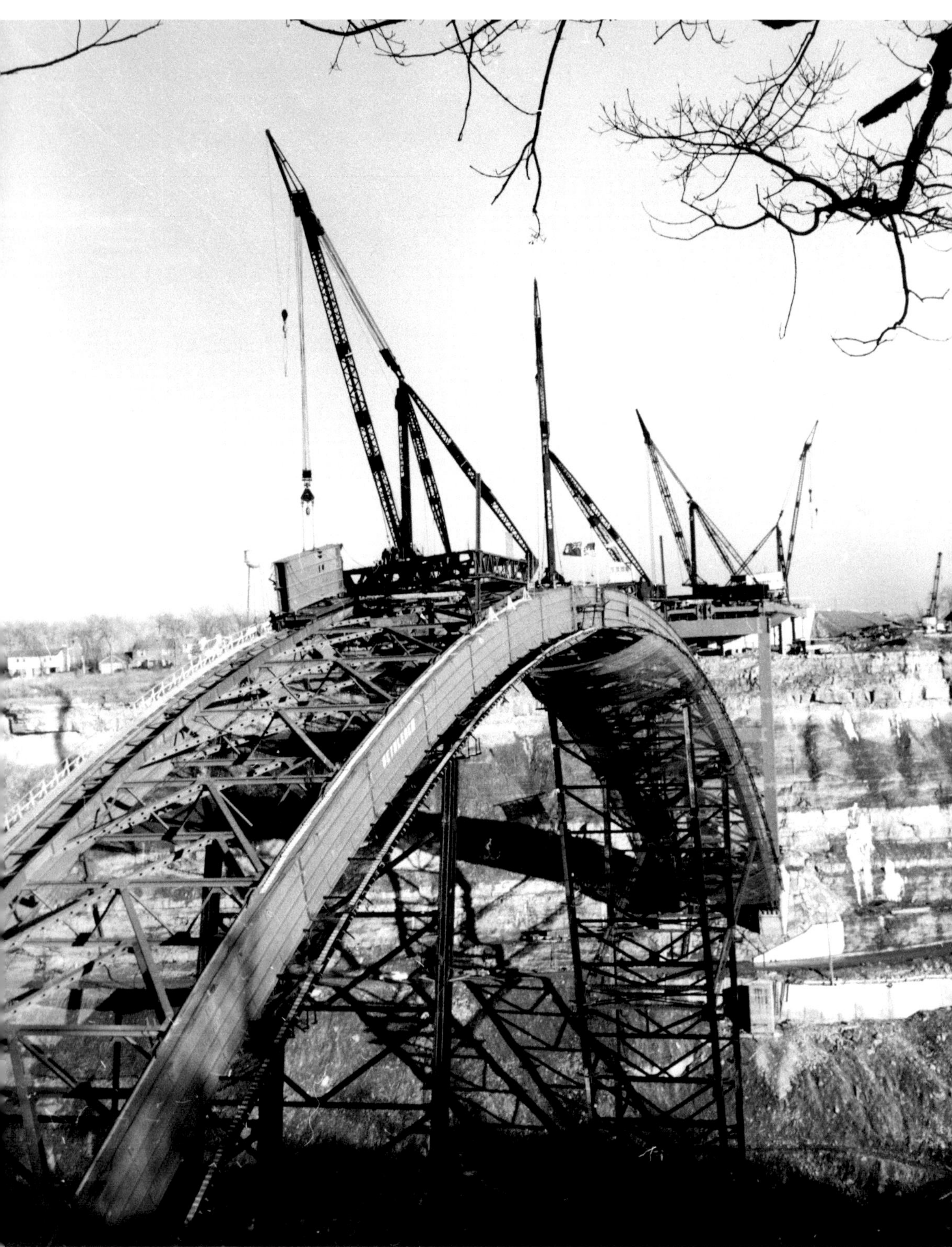

The Lewiston-Queenston Bridge under construction, 1962. (*H&H*)

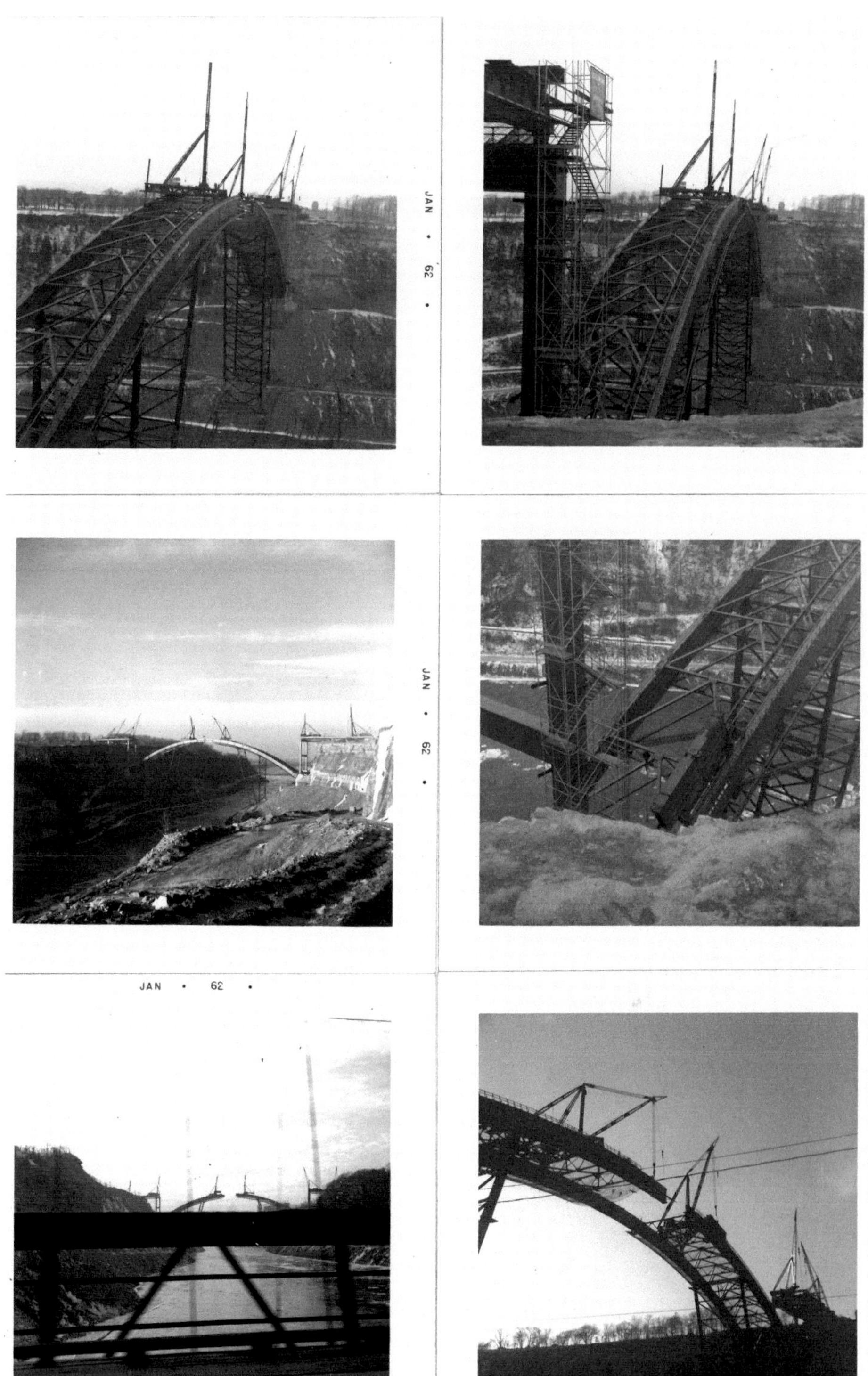

Six photos of construction on the Lewiston-Queenston Bridge in January 1962. (*H&H*)

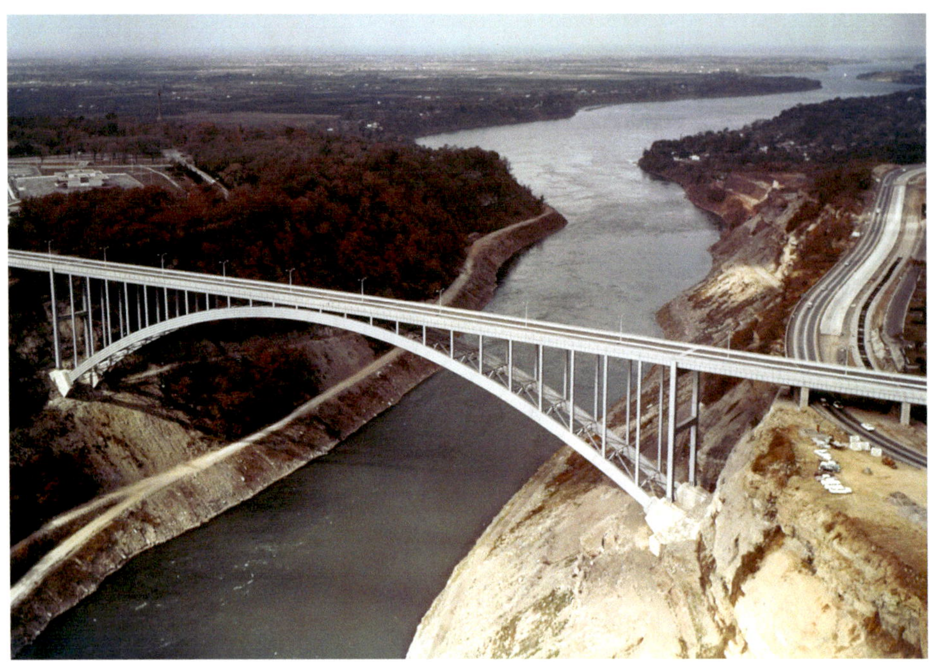

A color image of the recently completed Lewiston-Queenston Bridge, *c.* 1962. (*H&H*)

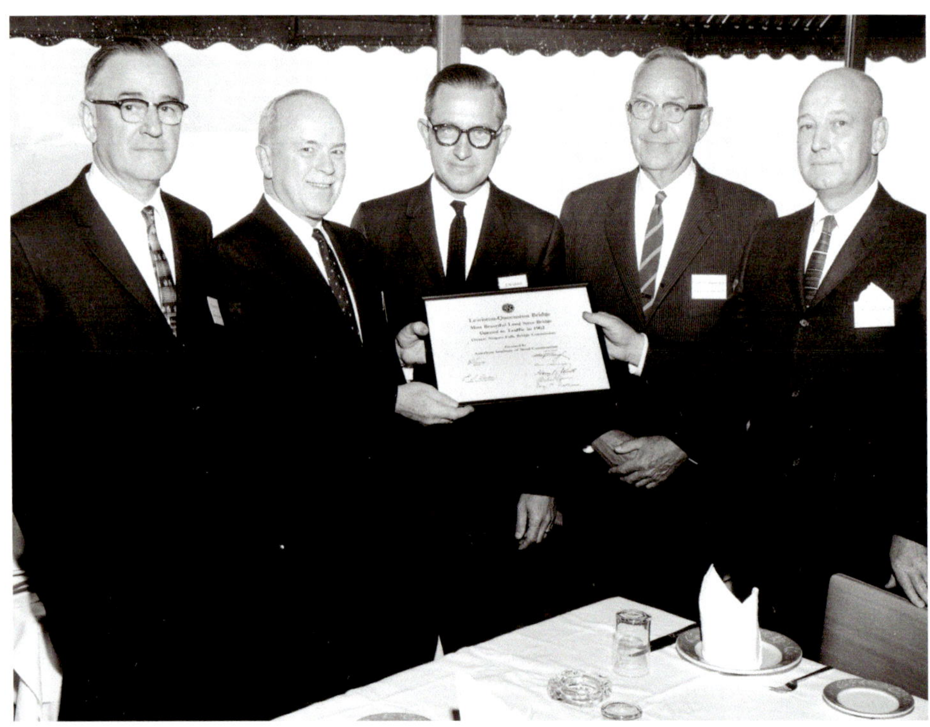

A group of officials accepts the 1962 "Most Beautiful Long Span Bridge Opened to Traffic in 1962" award from the American Institute of Steel Construction. (*H&H*)

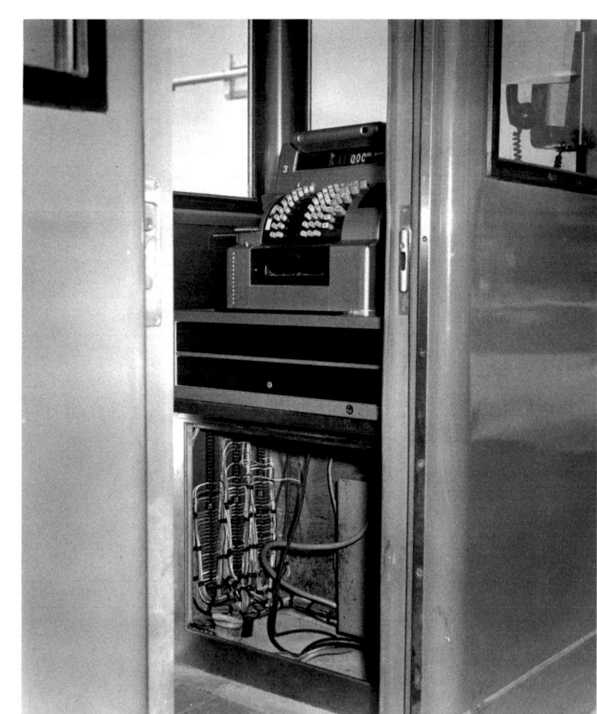

Right: Toll equipment at the Lewiston Queenston Bridge in June 1963, including a National brand cash register. (*H&H*)

Below: The Lewiston-Queenston Bridge toll plaza, 1963. (*H&H*)

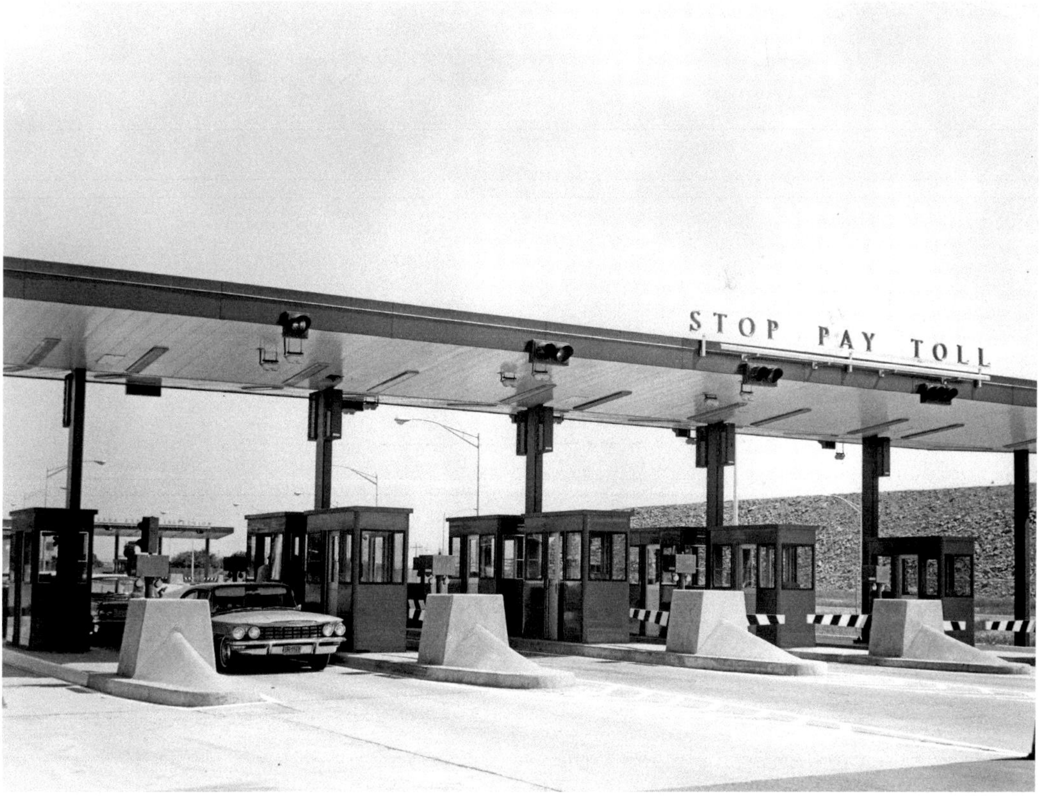

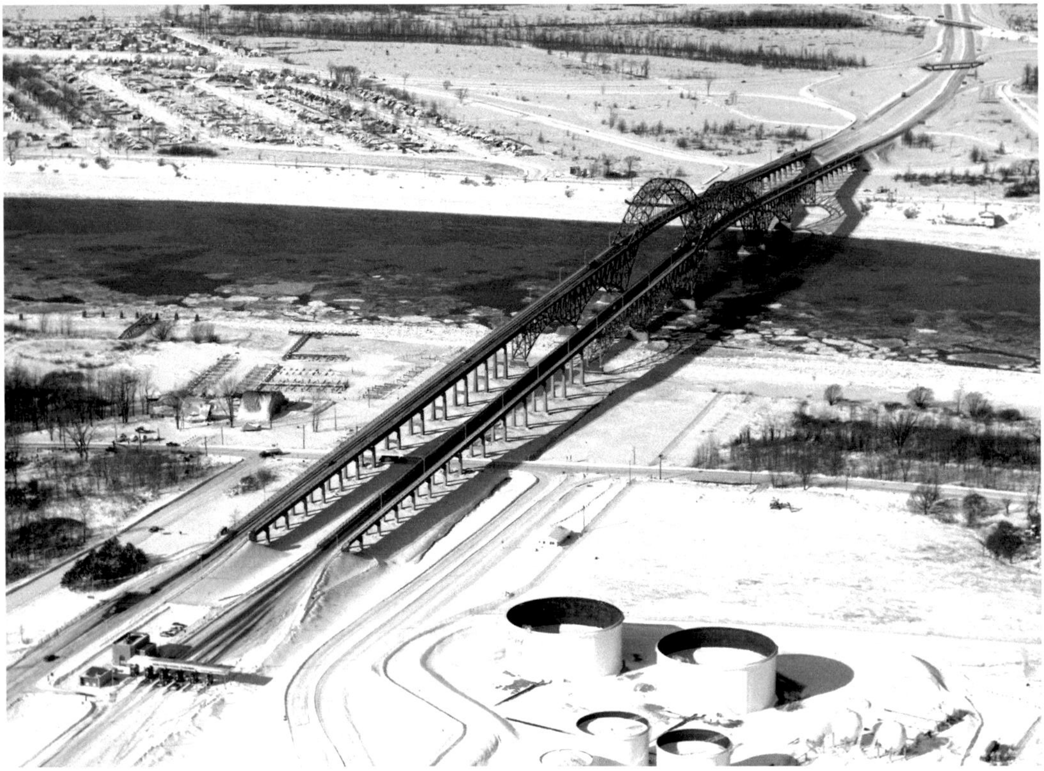

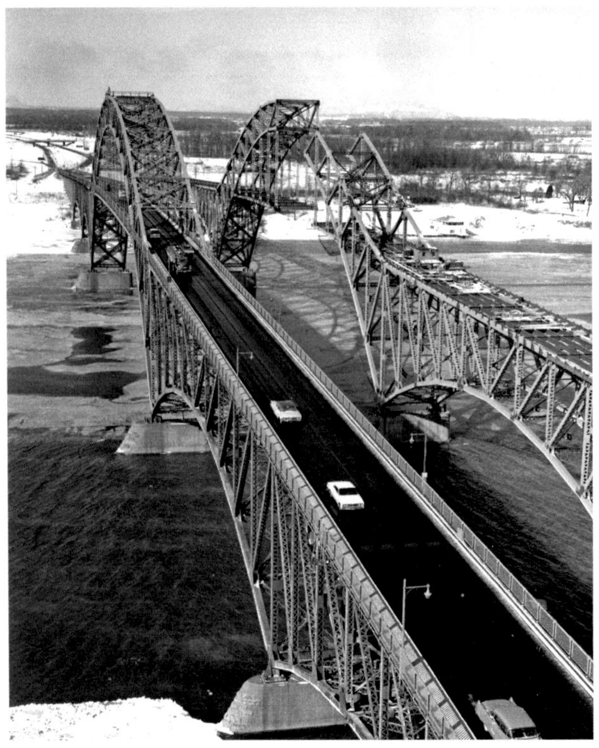

Above: The twin South Grand Island bridges over the Niagara River as seen in a snowy scene in 1963. (*H&H*)

Left: The South Grand Island Bridge under construction in 1963 parallel to the original 1935 bridge. In this photograph, the steel arch and central portion of the bridge deck are almost complete. (*H&H*)

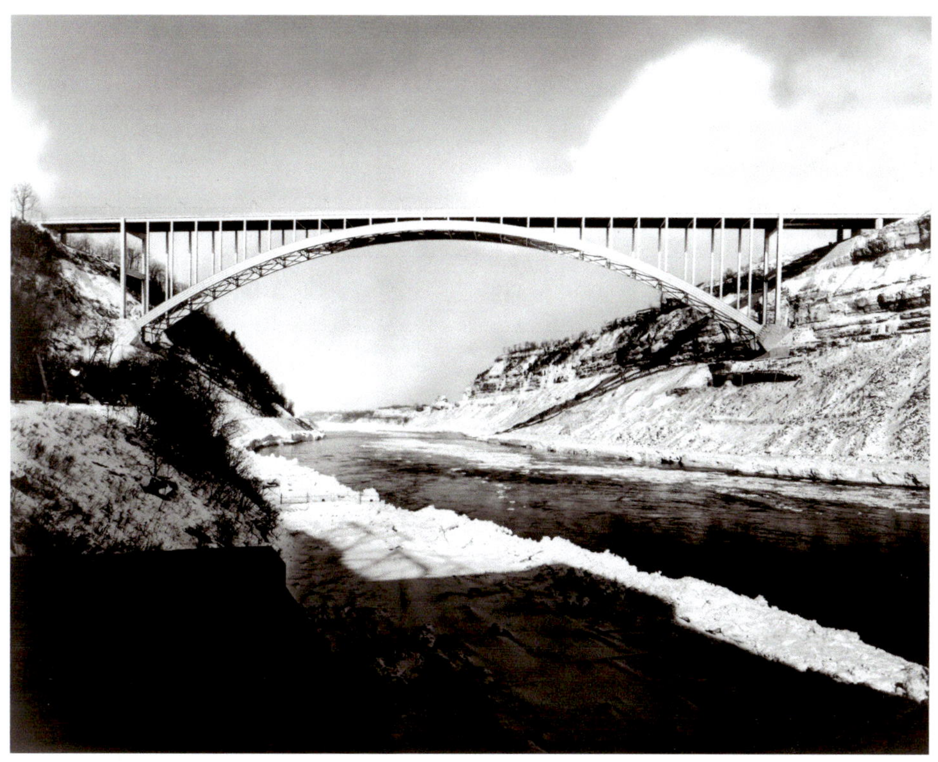

A view of the completed Lewiston-Queenston Bridge in winter early *c.* 1960s, with the ice and snow visible in the river and on the banks. (*H&H*)

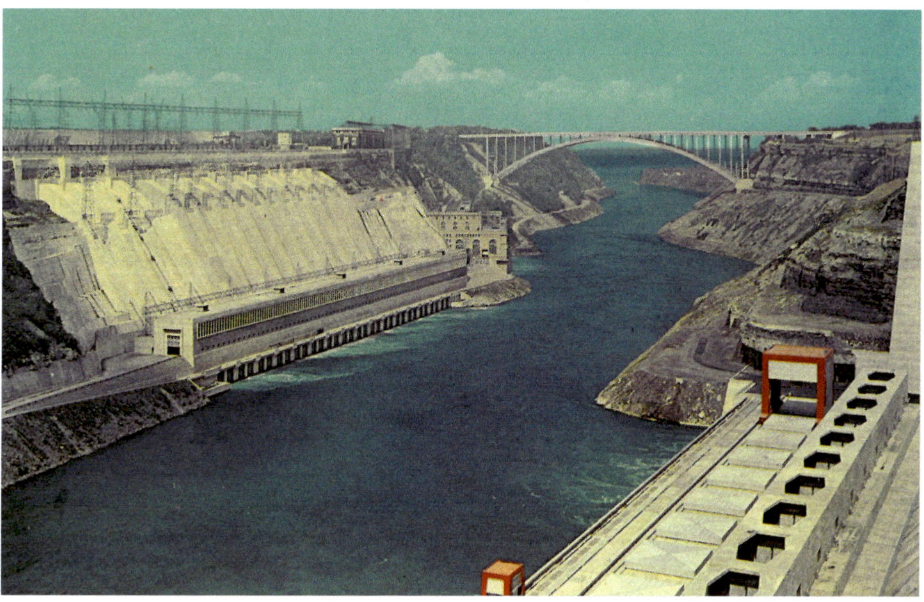

The Robert Moses Niagara Power Plant (left) and the Canadian Sir Adam Beck Generating Plant in a 1960s view, with the Lewiston-Queenston Bridge in the background. (*H&H*)

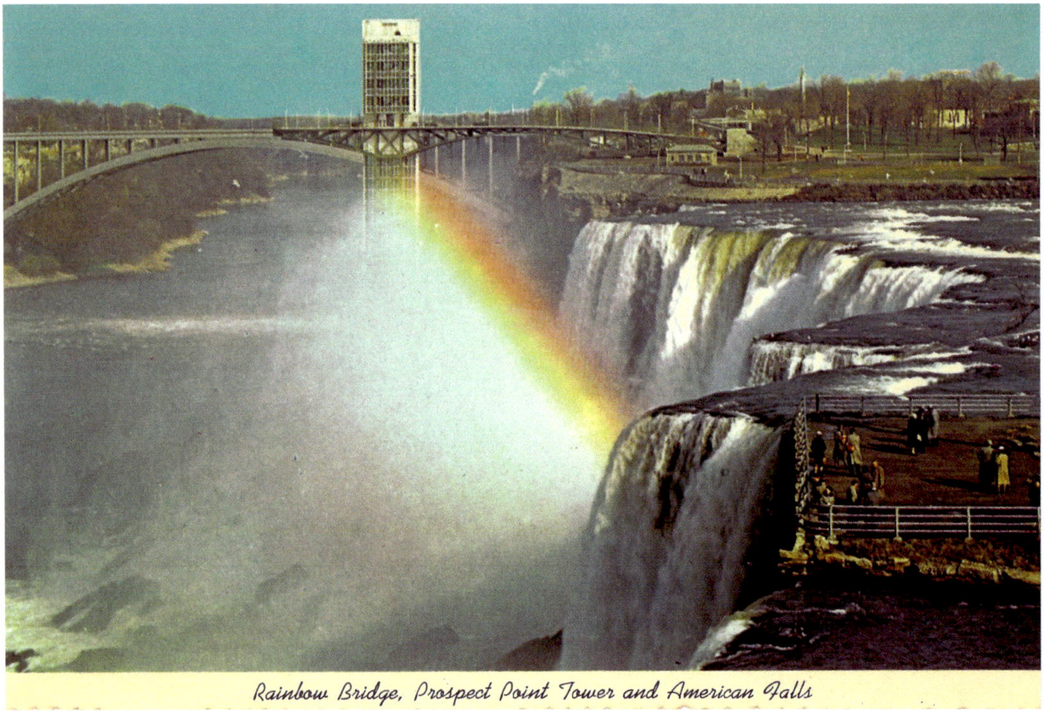

A 1960s view of the Falls, Rainbow Bridge, and the Niagara Falls Observation Tower. An elevator from the tower goes down to the *Maid of the Mist* dock.

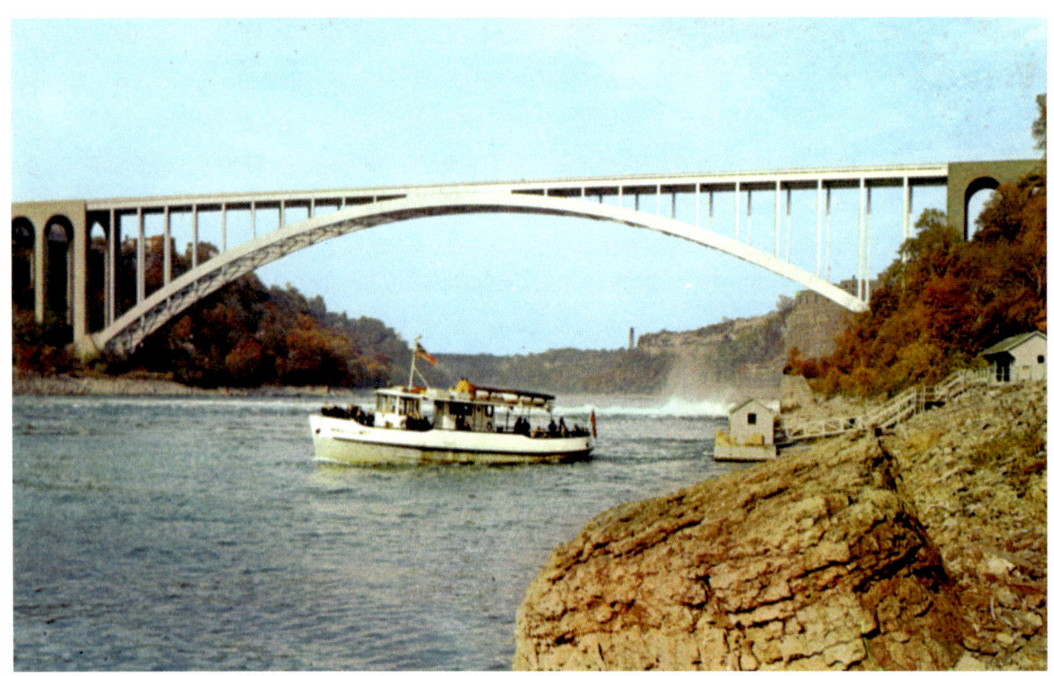

The *Maid of the Mist* near the Rainbow Bridge, *c.* 1960s.

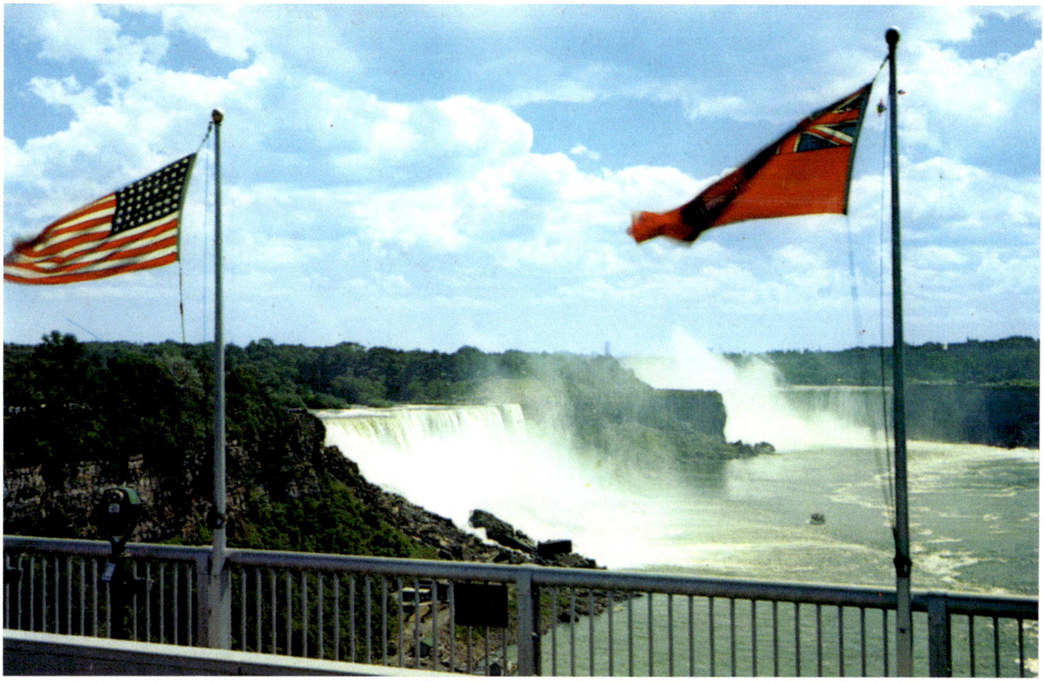

View of Niagara Falls taken from the border point on the Rainbow Bridge, *c.* 1960s.

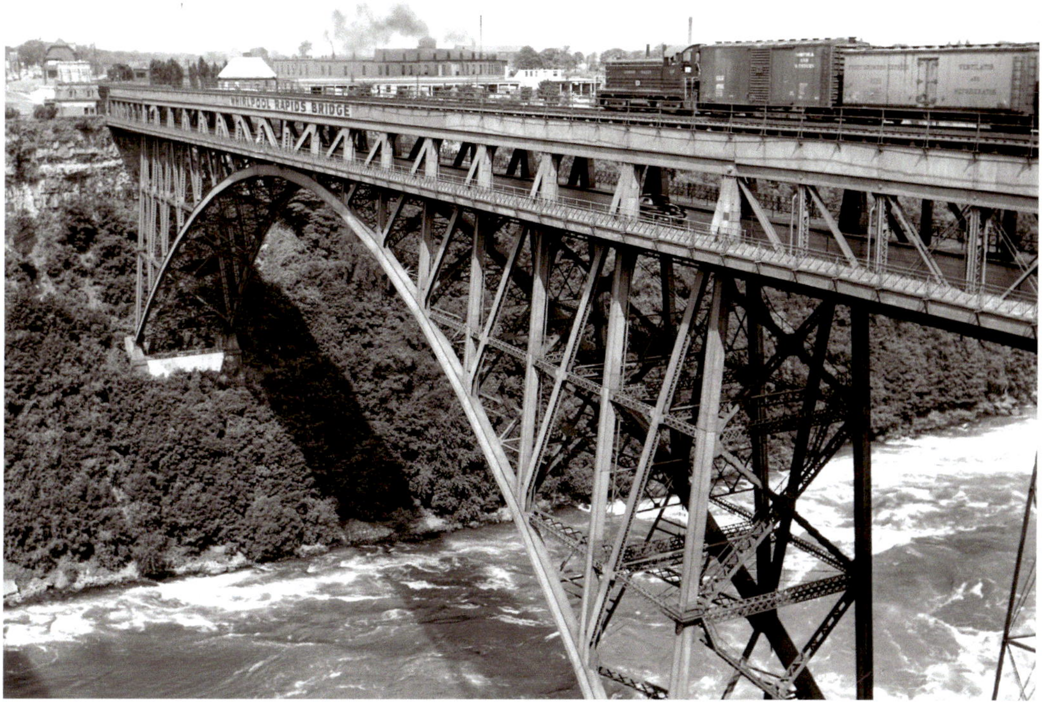

The Whirlpool Rapids Bridge by Niagara Falls photographer Morey Englander, *c.* 1960s. (*H&H*)

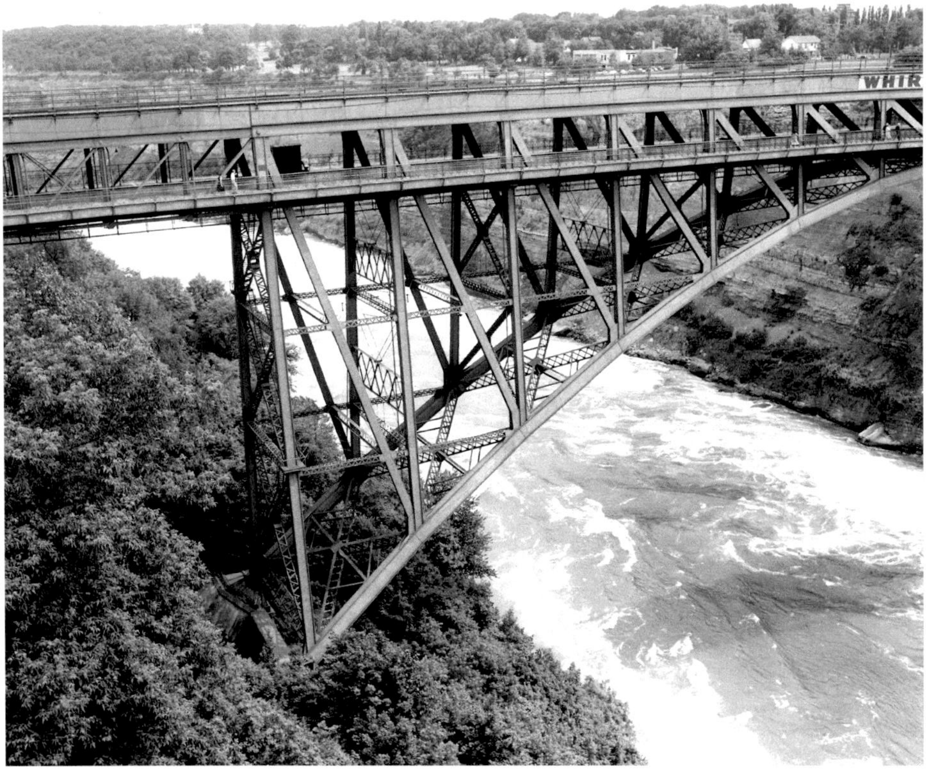

Above and below: Photos of the Whirlpool Rapids Bridge by Niagara Falls photographer Morey Englander, *c.* 1960s. (*H&H*)

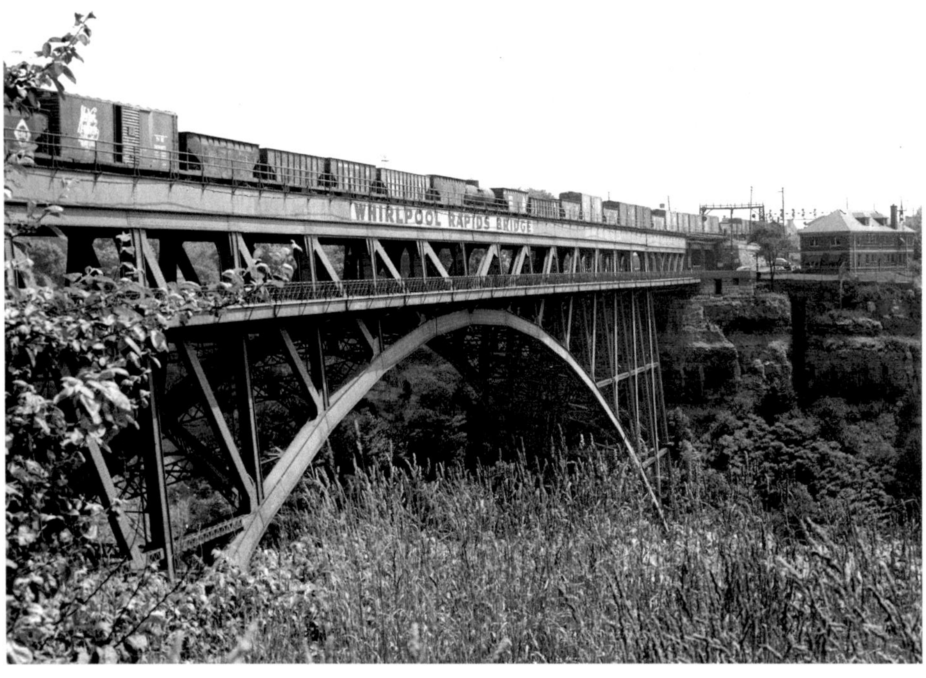

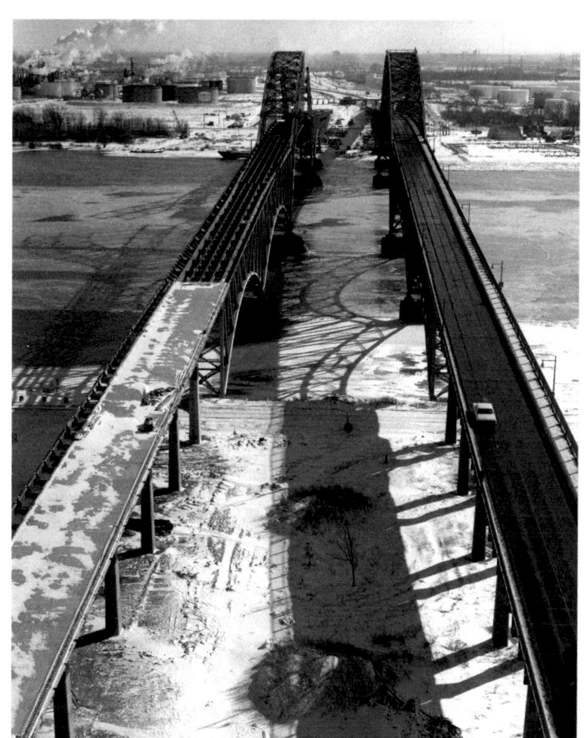

Right: The South Grand Island Bridge under construction in 1963. (*H&H*)

Below: The North Grand Island Bridges, *c.* 1960s. (*H&H*)

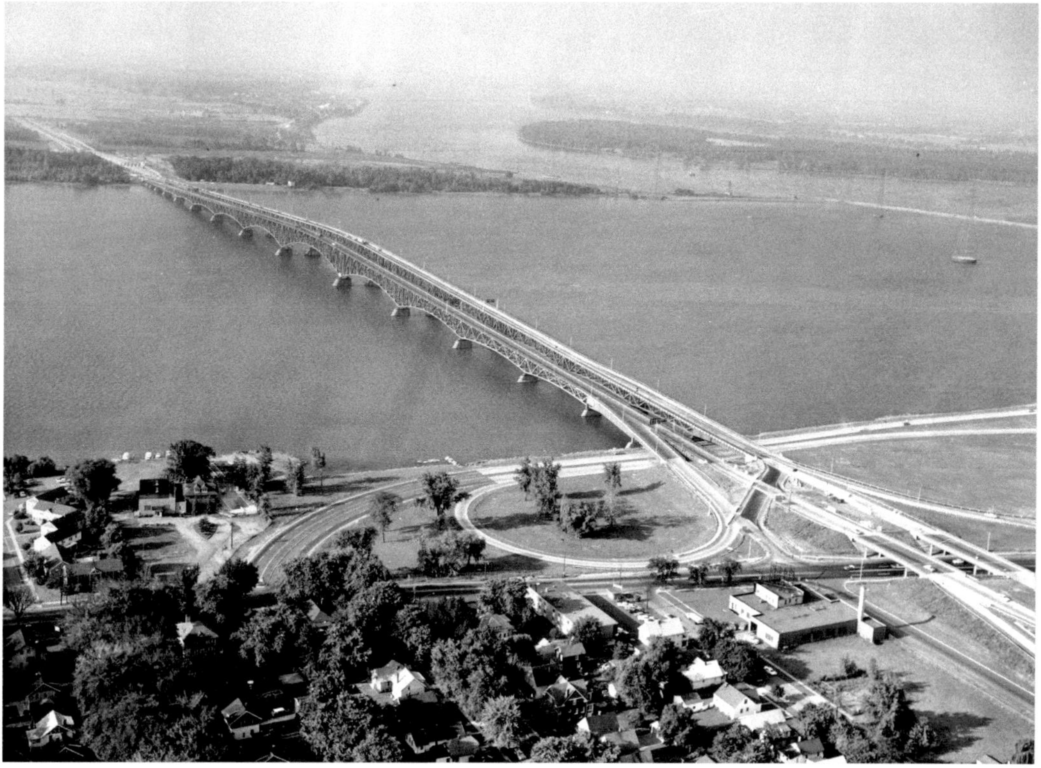

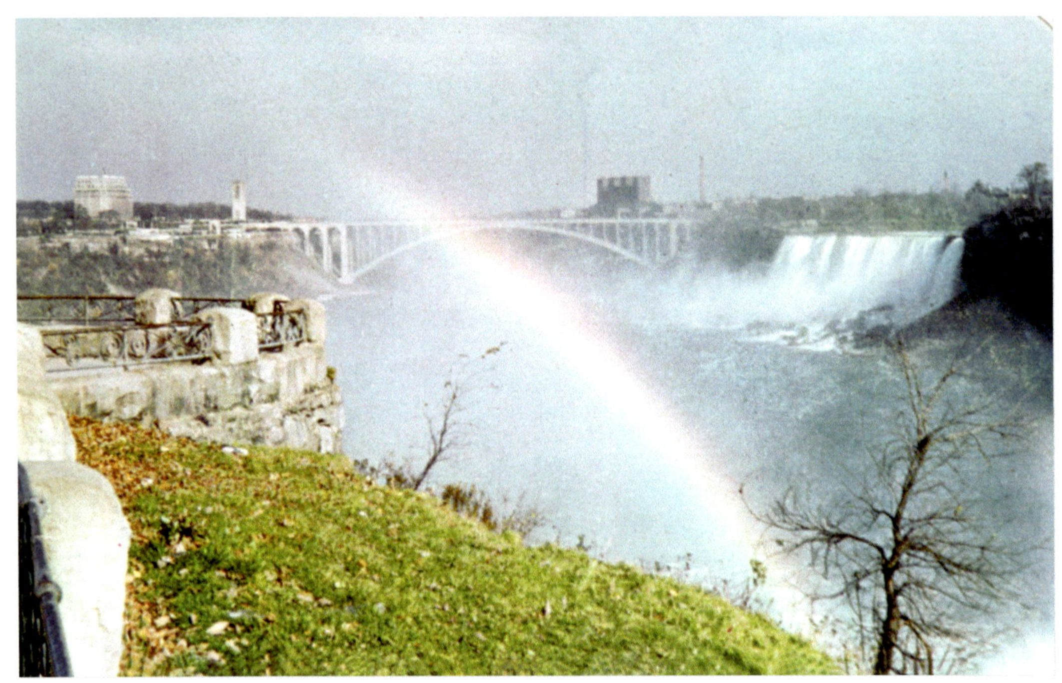

Above and below: Rainbow Bridge (faintly visible in the background) is aptly named as these *c.* 1960s postcard images show. The caption for the one at top reads: "The ever present mist at Niagara Falls causes a multitude of varied colored rainbows." The bottom postcard caption reads: "On sunny days, the year round, beautiful rainbows hover above the falls, caused by the ever present spray from the tumbling waters."

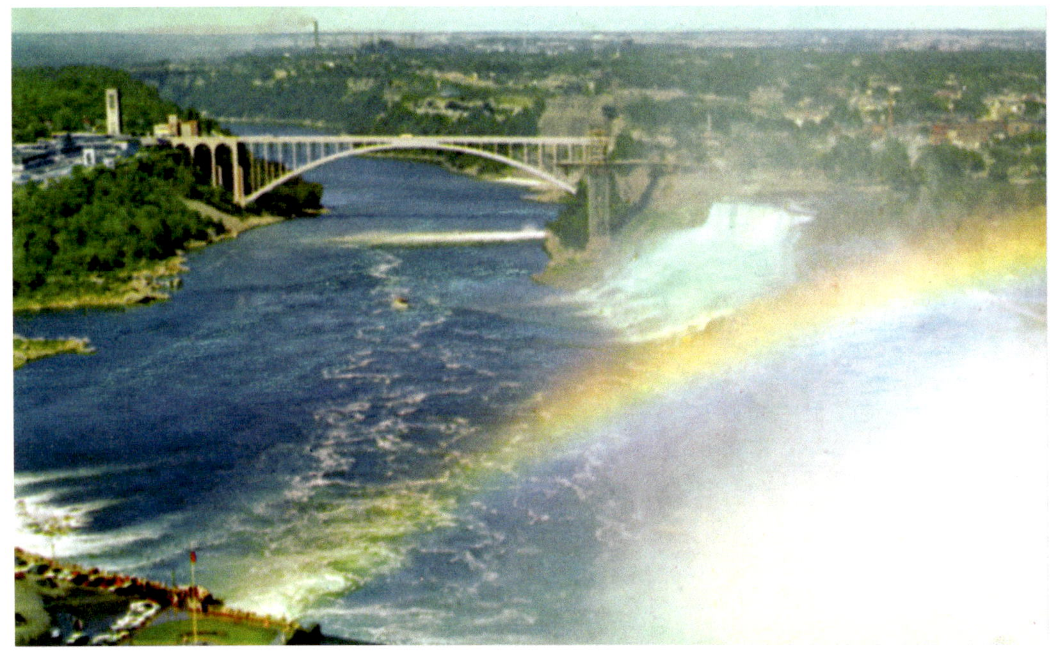

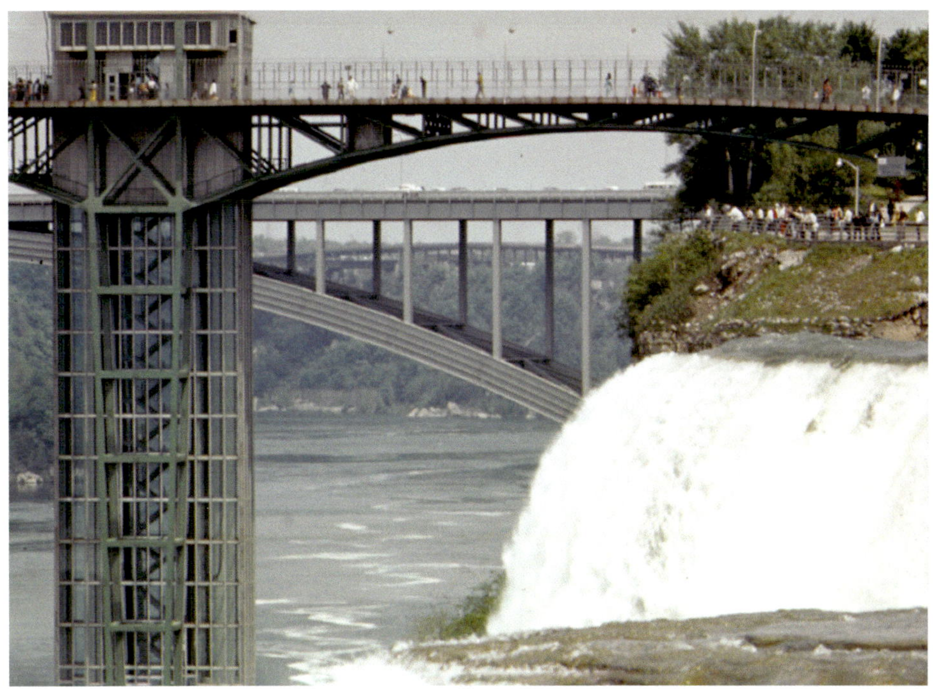

A view of the American Falls from Goat Island, showing a portion of the Prospect Point Observation Tower and Rainbow Bridge in the background, 1973. (*National Archives*)

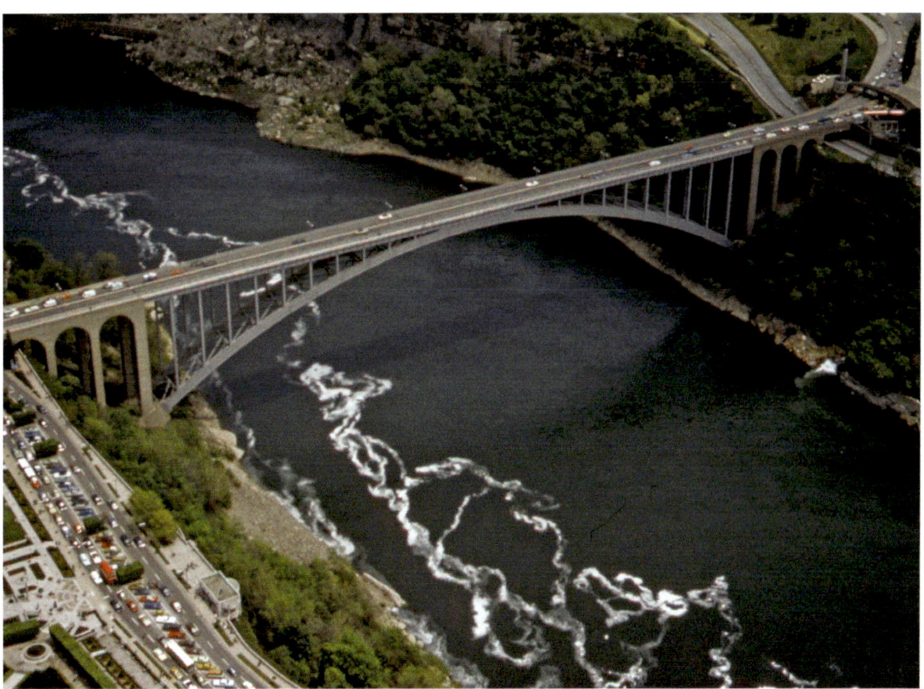

Aerial view of the Niagara River below the falls, showing the Rainbow Bridge, 1973. (*National Archives*)

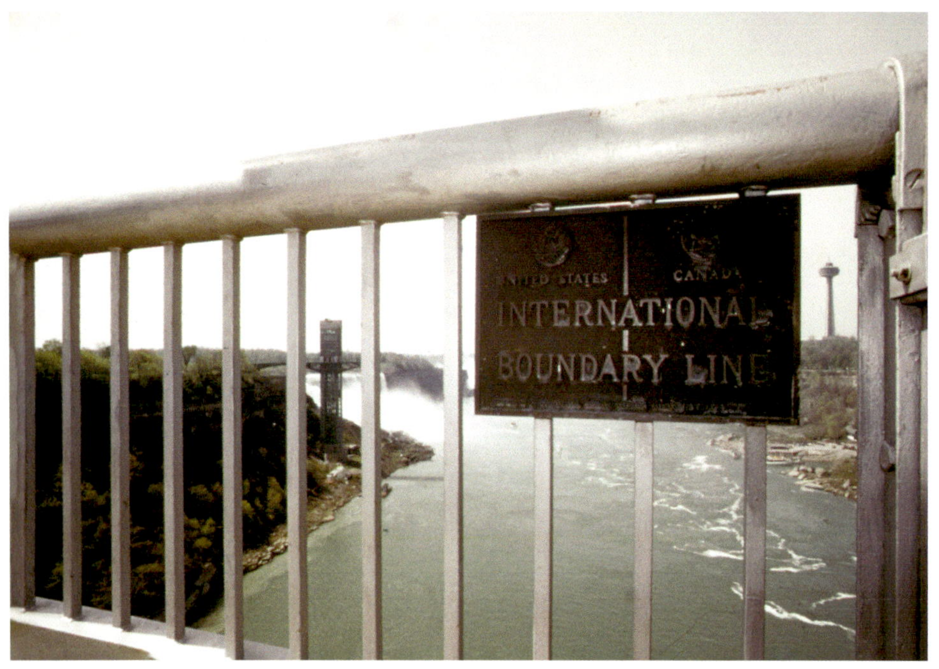

A sign on the guiderail of the Rainbow Bridge in this 1973 photo marks the boundary line between Canada and the United States. The American Falls are visible through the railings. (*National Archives*)

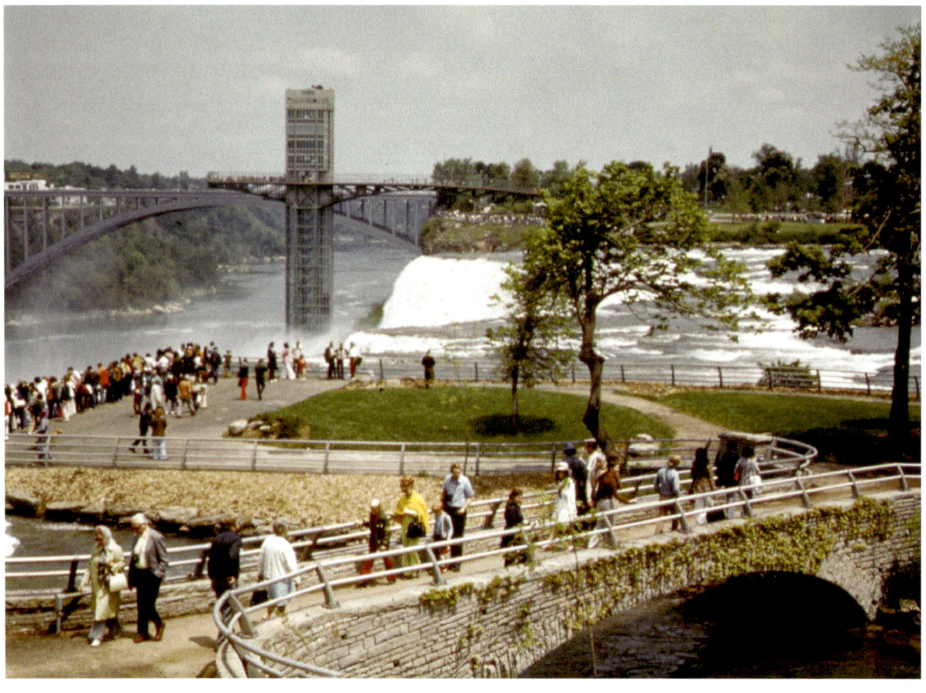

Luna Island on the American side of the Niagara River looking toward the American Falls and Observation Tower with the Rainbow Bridge in the background, 1973. (*National Archives*)

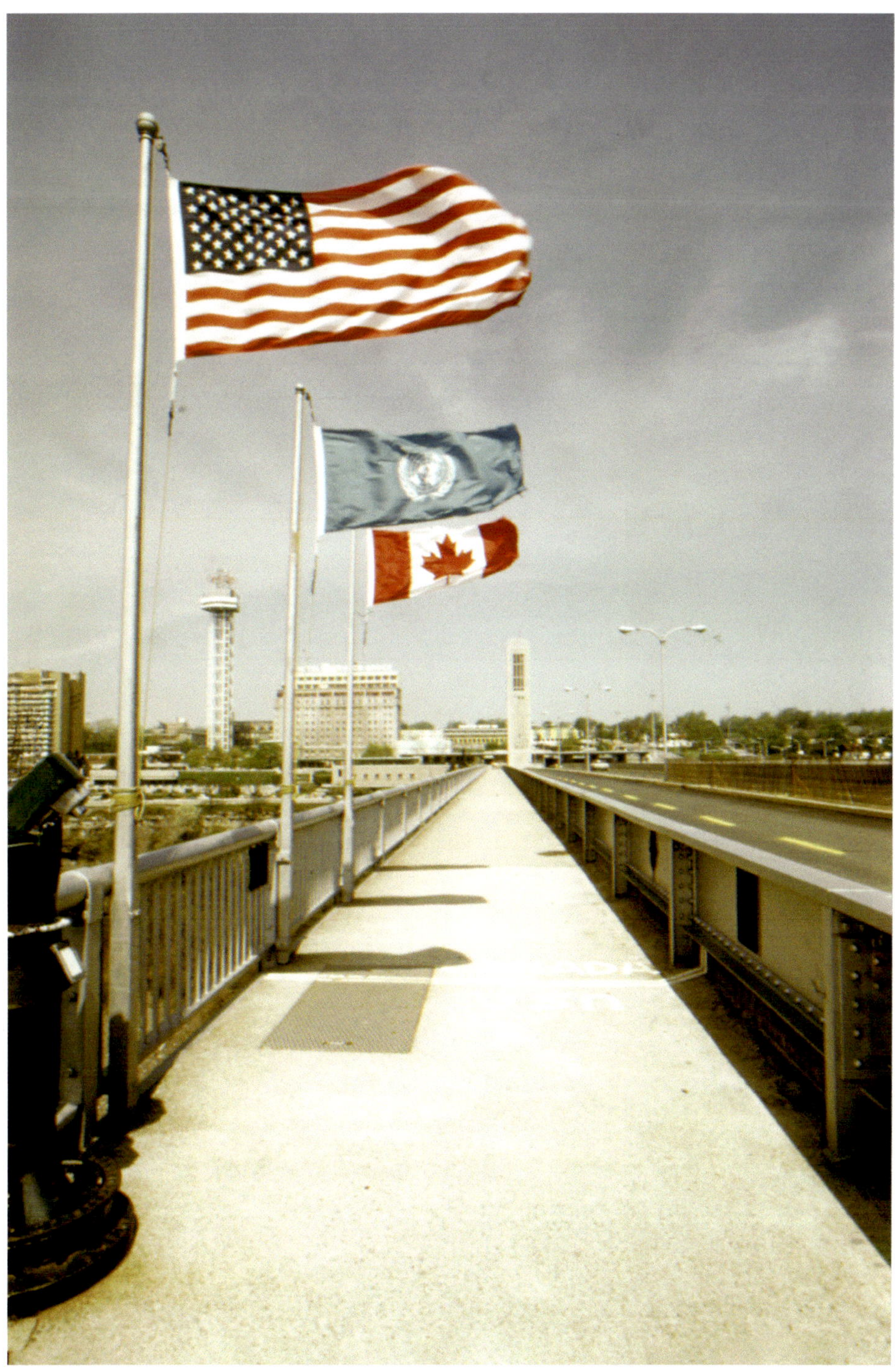

This 1973 photograph, taken from the deck of the Rainbow Bridge, shows the Carillon Bell Tower at the Ontario bridge plaza, and the thirteen-story Hotel General Brock. Built in 1929, it was the first high-rise hotel in Niagara Falls. Marilyn Monroe stayed at the hotel while filming the movie *Niagara*. (*National Archives*)

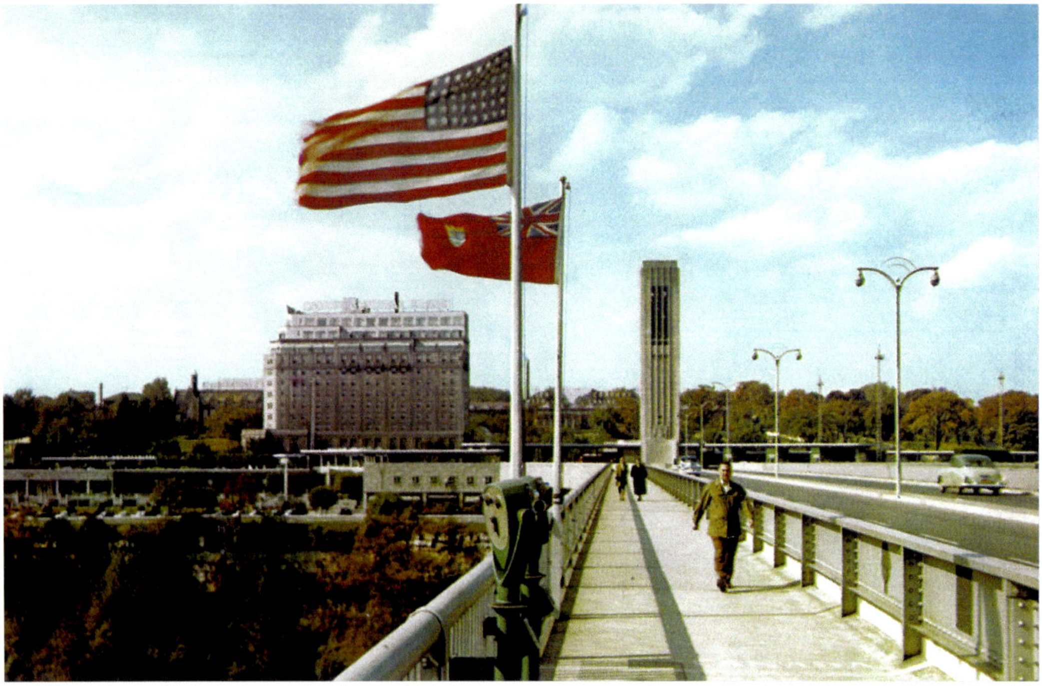

A *c.* 1970s postcard taken from the deck of the Rainbow Bridge.

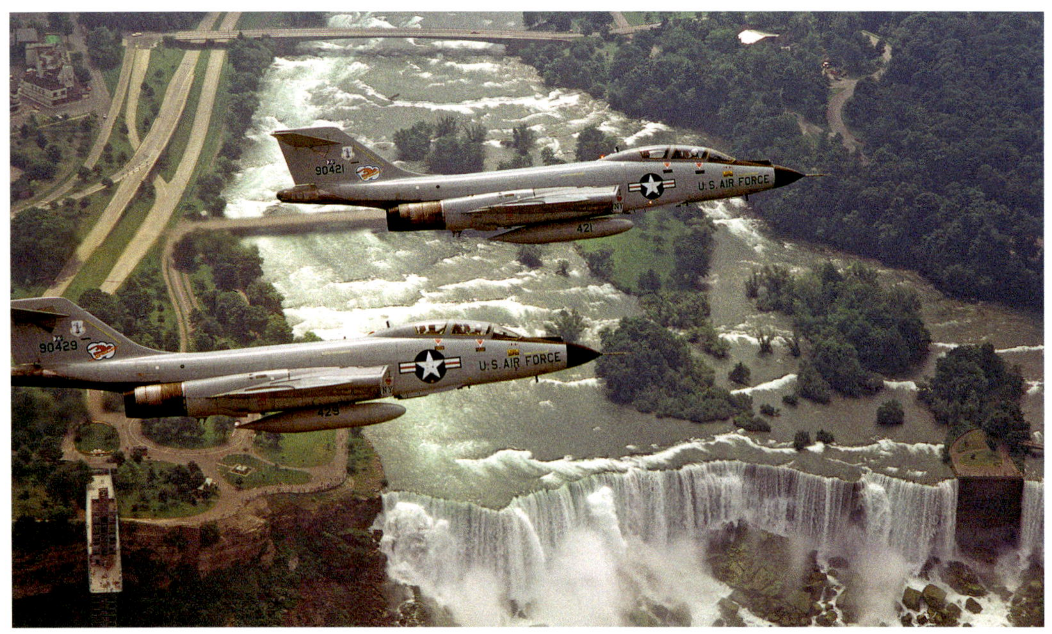

An air-to-air right-side view of two F-101 Voodoo aircraft over Niagara Falls during exercise Sentry Castle '81. The aircraft are assigned to the 107th Fighter Interceptor Group, 136th Fighter Interceptor Squadron, New York Air National Guard, July 1, 1981. The North Grand Island Bridge is in the background. (*National Archives*)

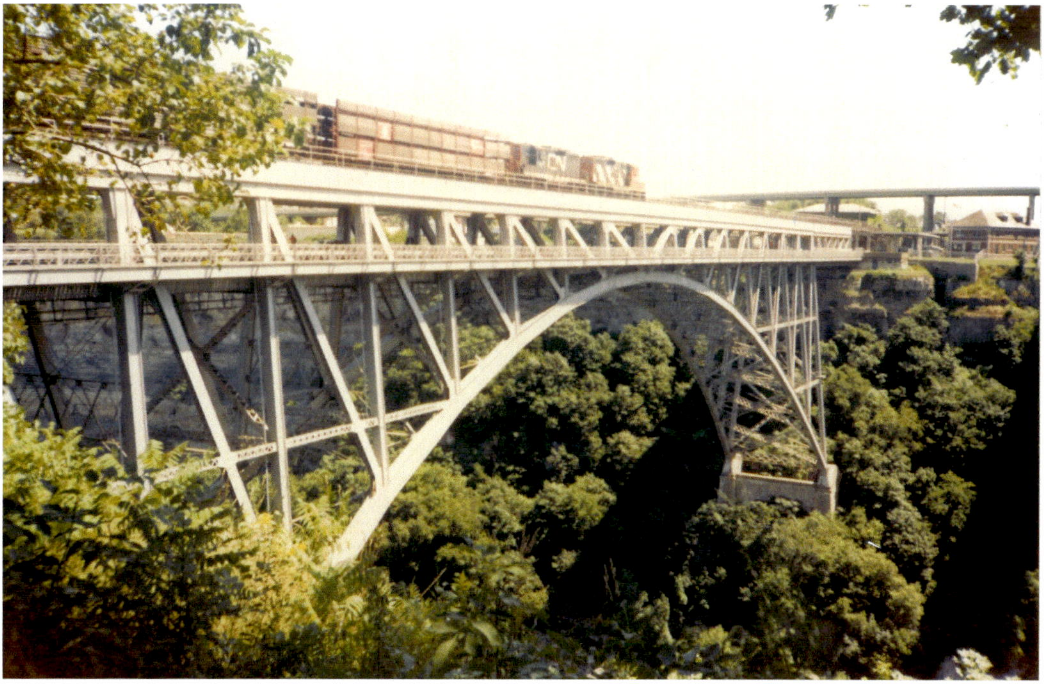

Whirlpool Rapids Bridge, 1981. (*H&H*)

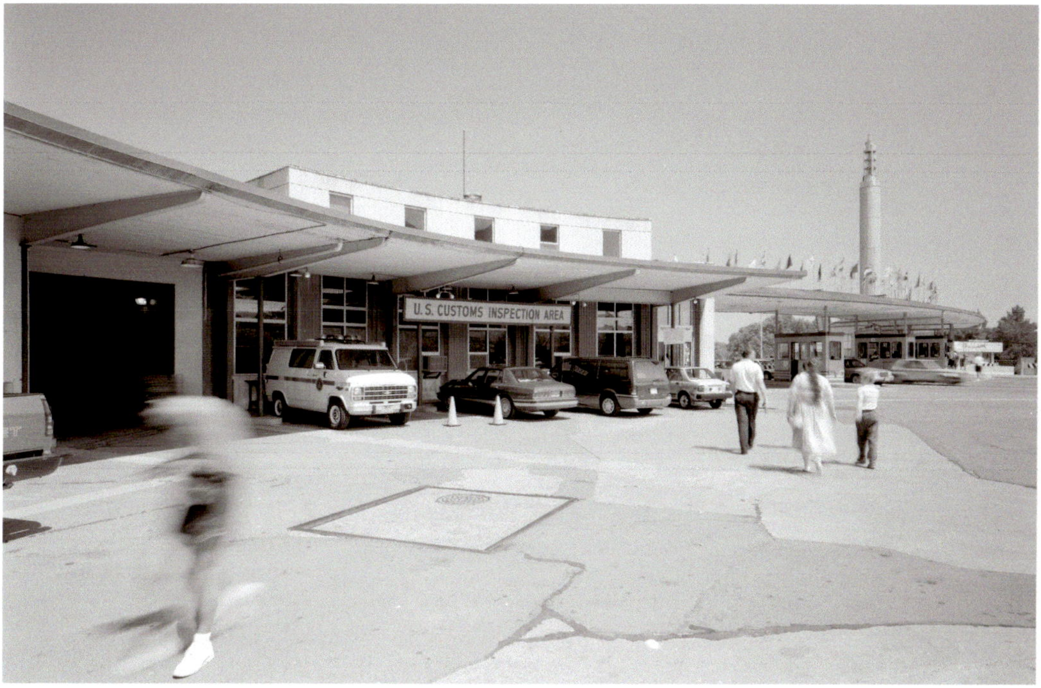

A 1994 Historic Buildings of America Survey photograph of the Rainbow Bridge Toll Plaza. (*National Archives*)

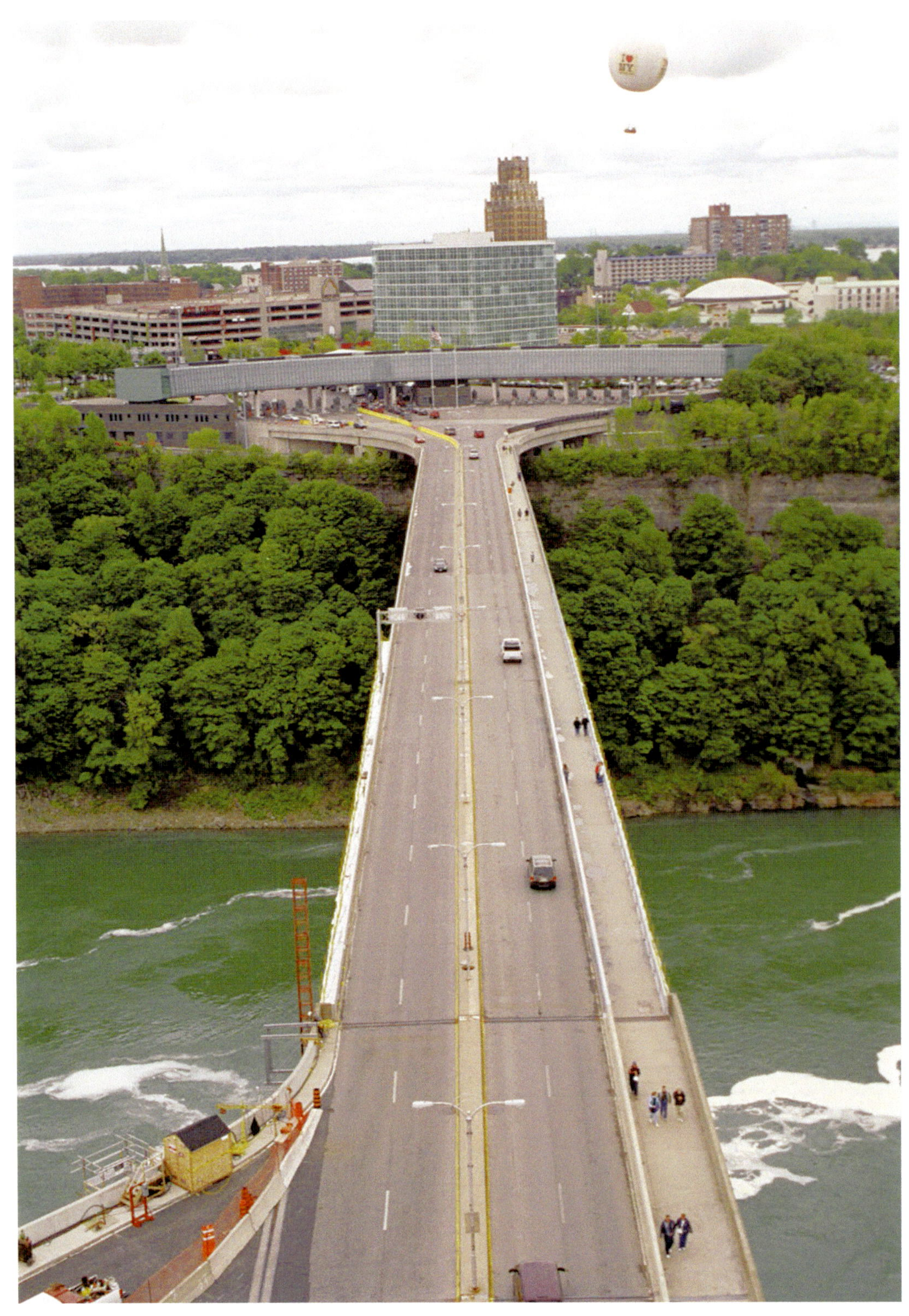

An aerial view of the Rainbow Bridge deck, *c.* 2014. (*H&H*)

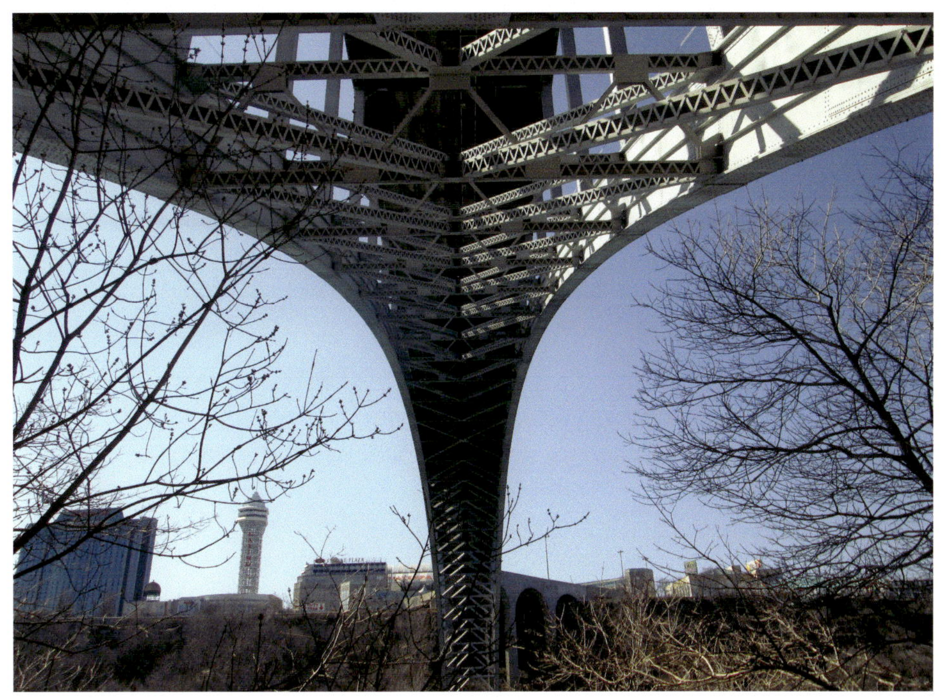

A view of the underside of the Rainbow Bridge showing the steel arch structure. (*H&H*)

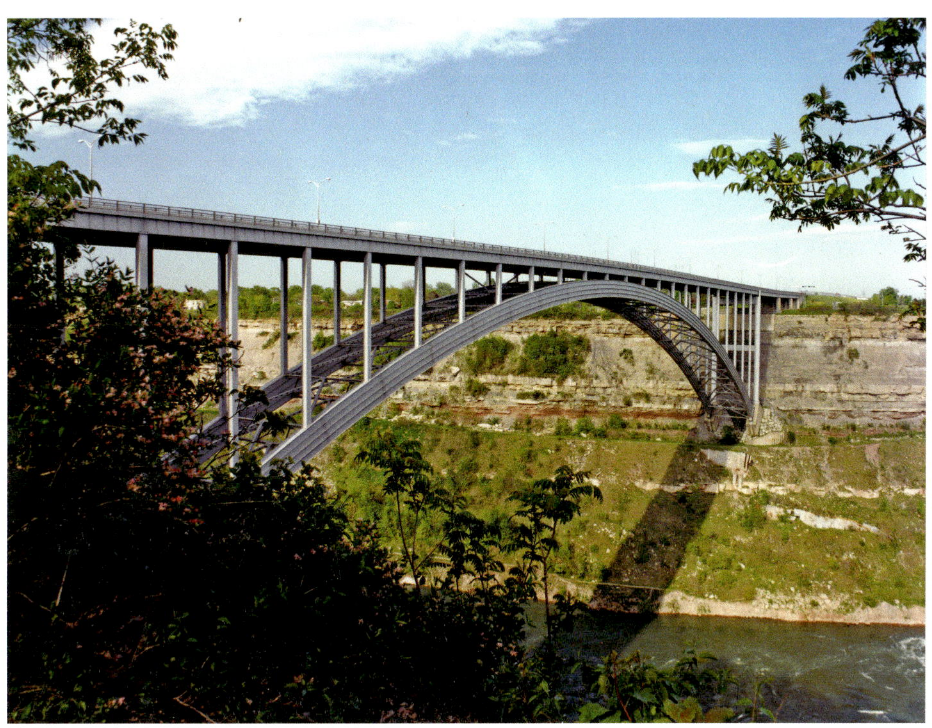

The Lewiston-Queenston Bridge in a recent photo. (*H&H*)

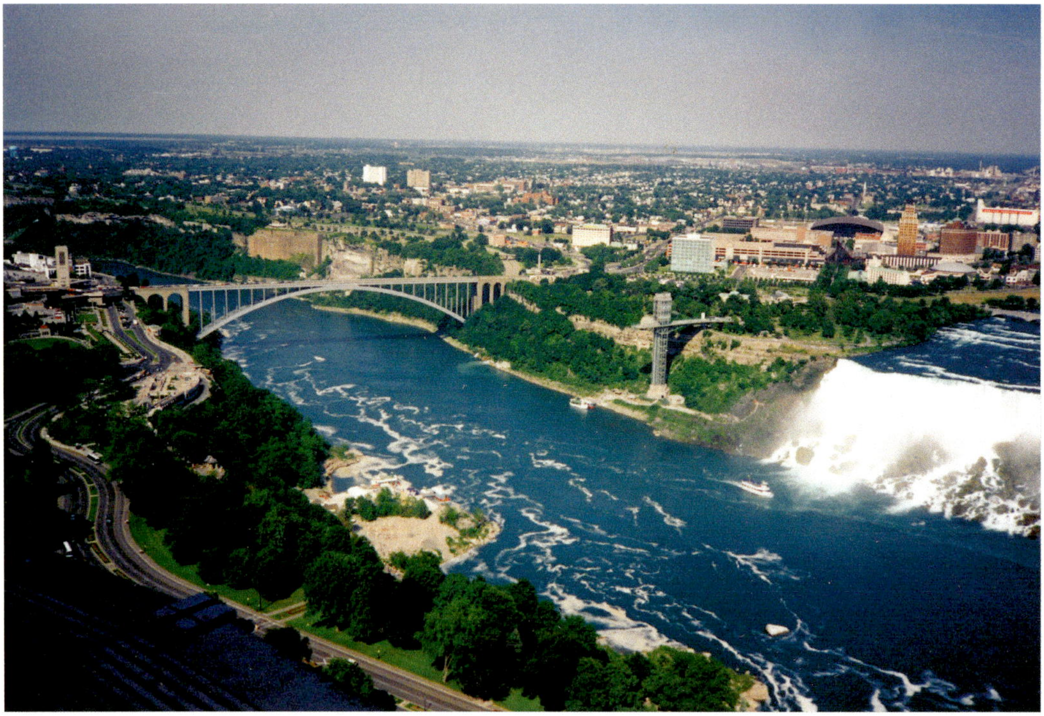

A scenic aerial view of the Rainbow Bridge. (*H&H*)

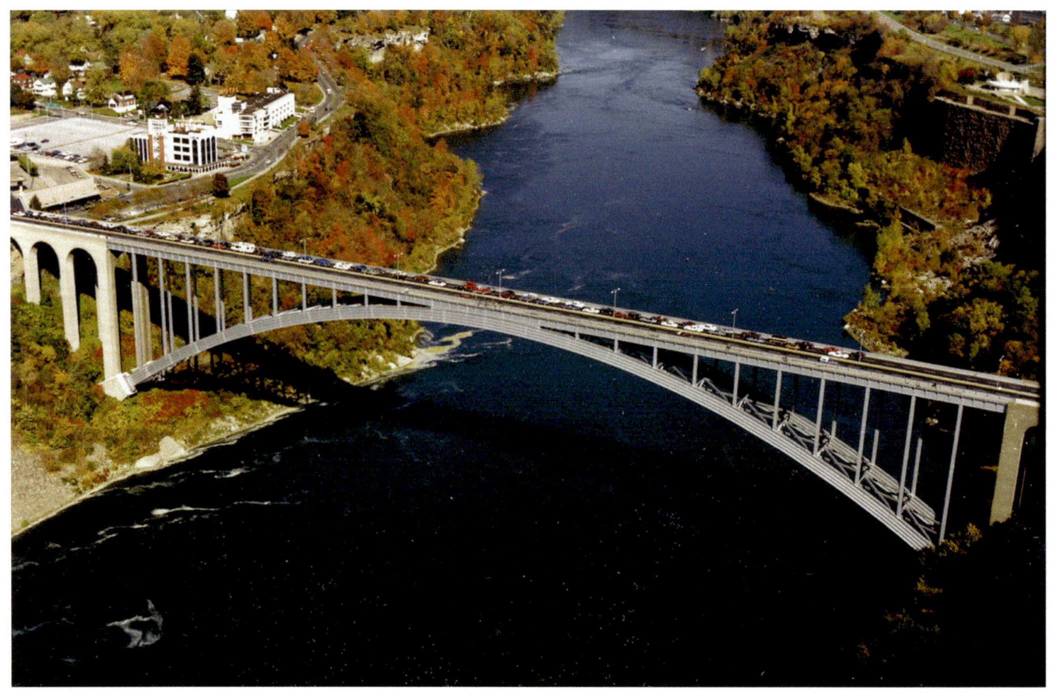

A scenic autumn aerial view of the Rainbow Bridge. (*H&H*)

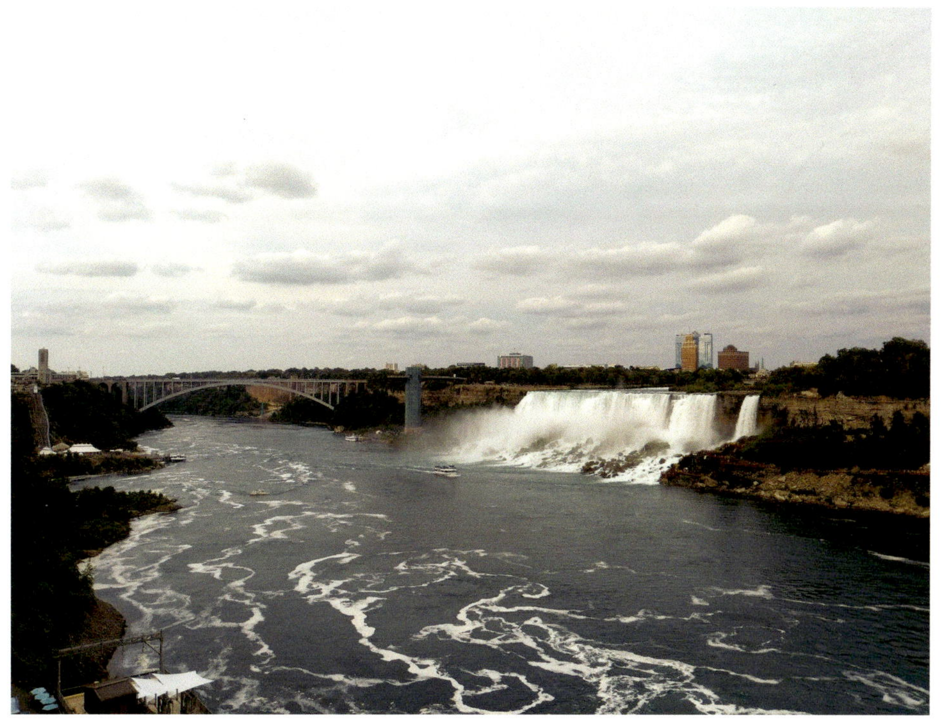

Above and below: Two 2018 photos of the Rainbow Bridge, an ubiquitous part of so many views of Niagara Falls since its completion in 1941.

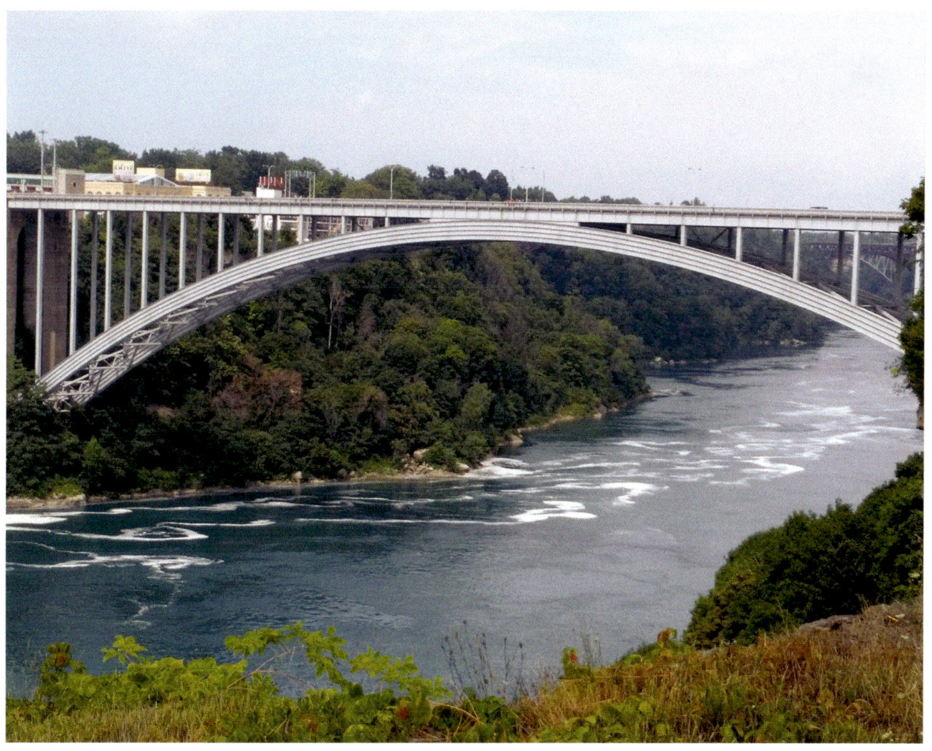

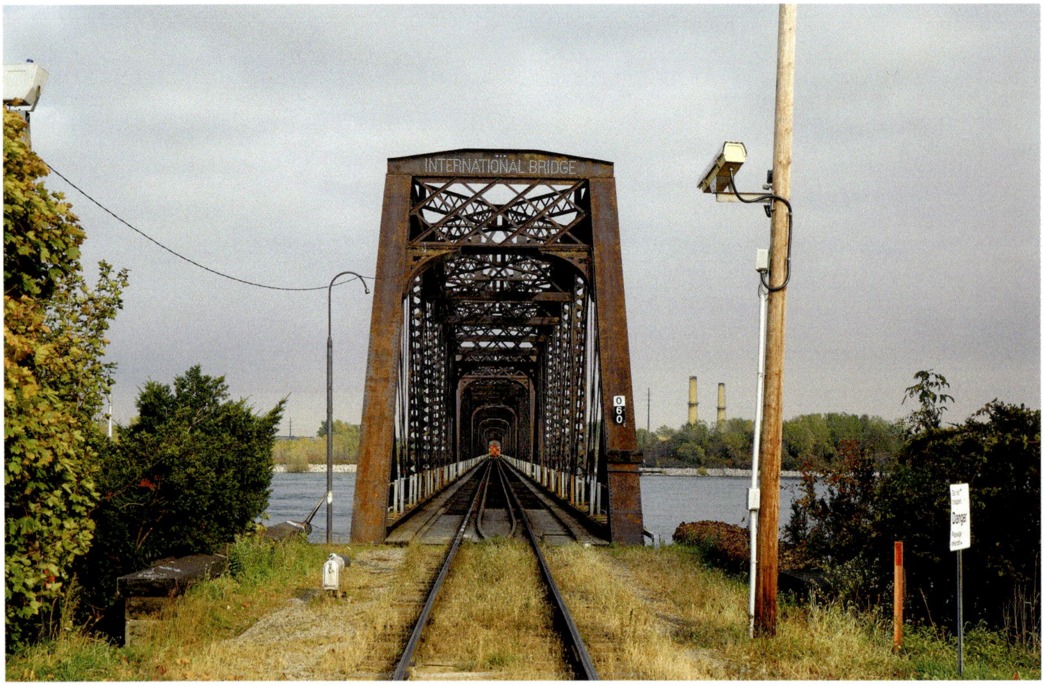

Above and below: The photo at top shows the Canadian portal of the International Bridge over the Niagara River looking east northeast toward Unity Island and Buffalo, and dates to 1992. The span was fixed in place during World War II. The photo at bottom is an aerial view of the bridge dating to 2018. (*Library of Congress*)

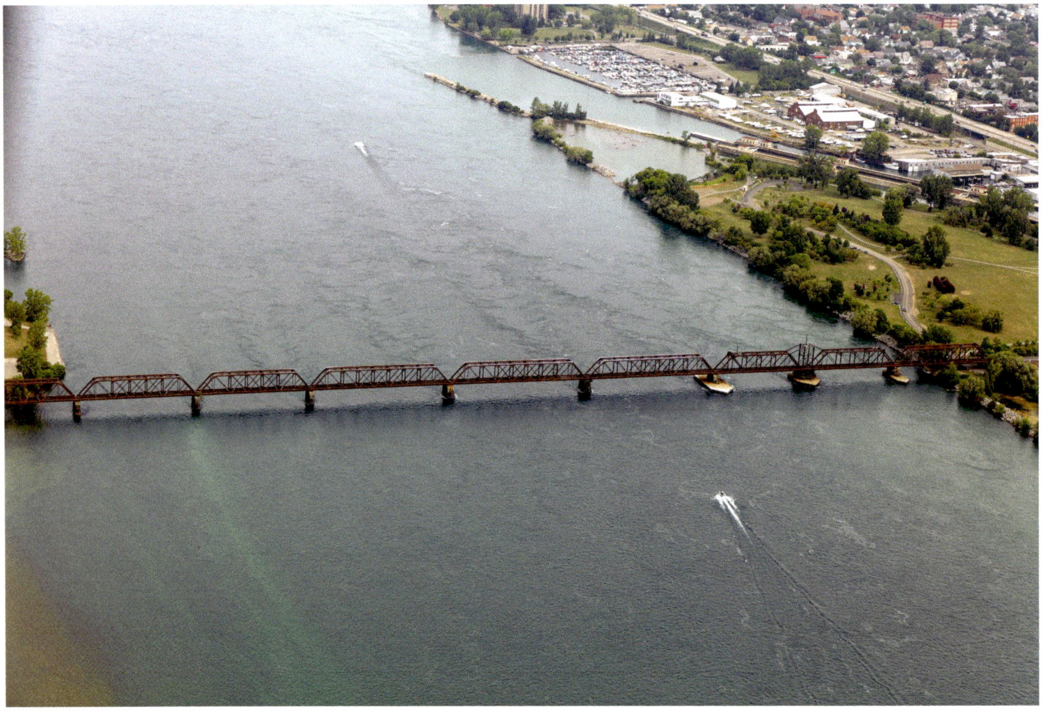

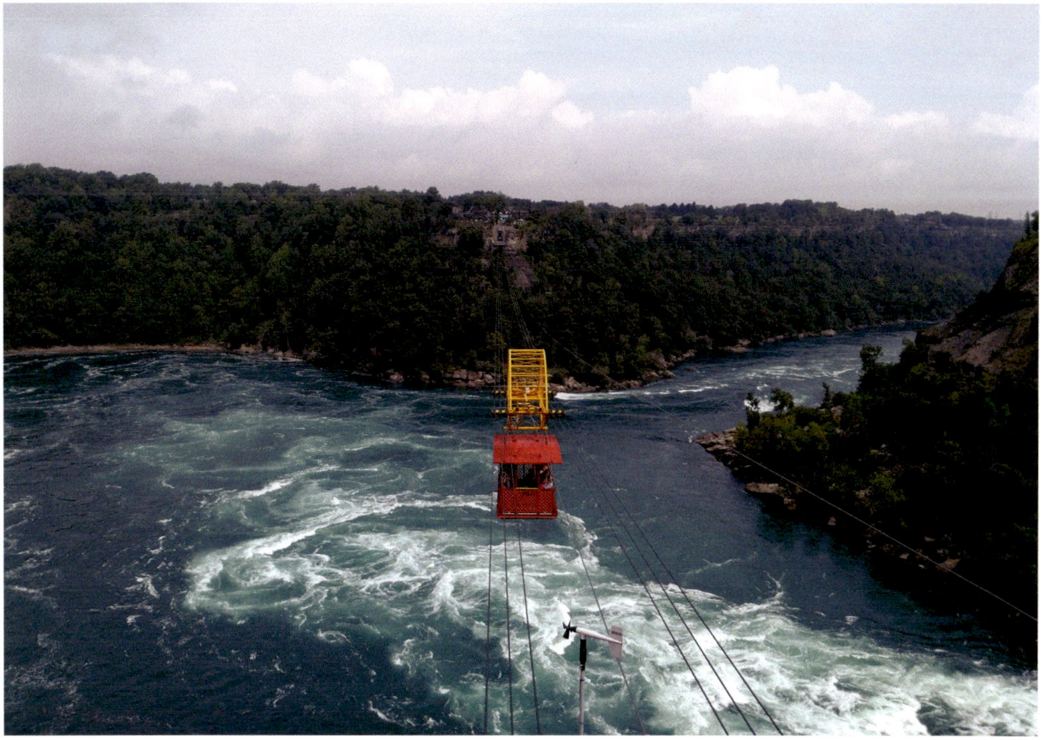

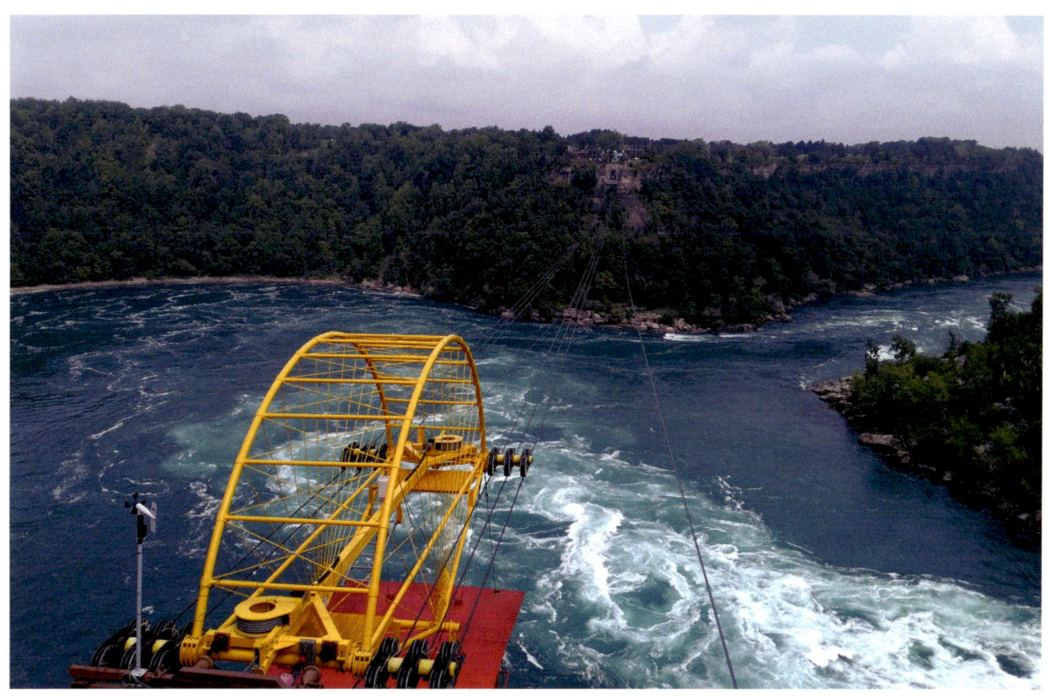

Above and below: The Whirlpool Aero Car (seen in 2018) is located on the Canadian side of the Whirlpool Rapids, a little bit north of Niagara Falls. The current iteration of the ride offers an experience that takes passengers 3,500 feet above the Niagara Gorge and the river's Class 6 rapids.

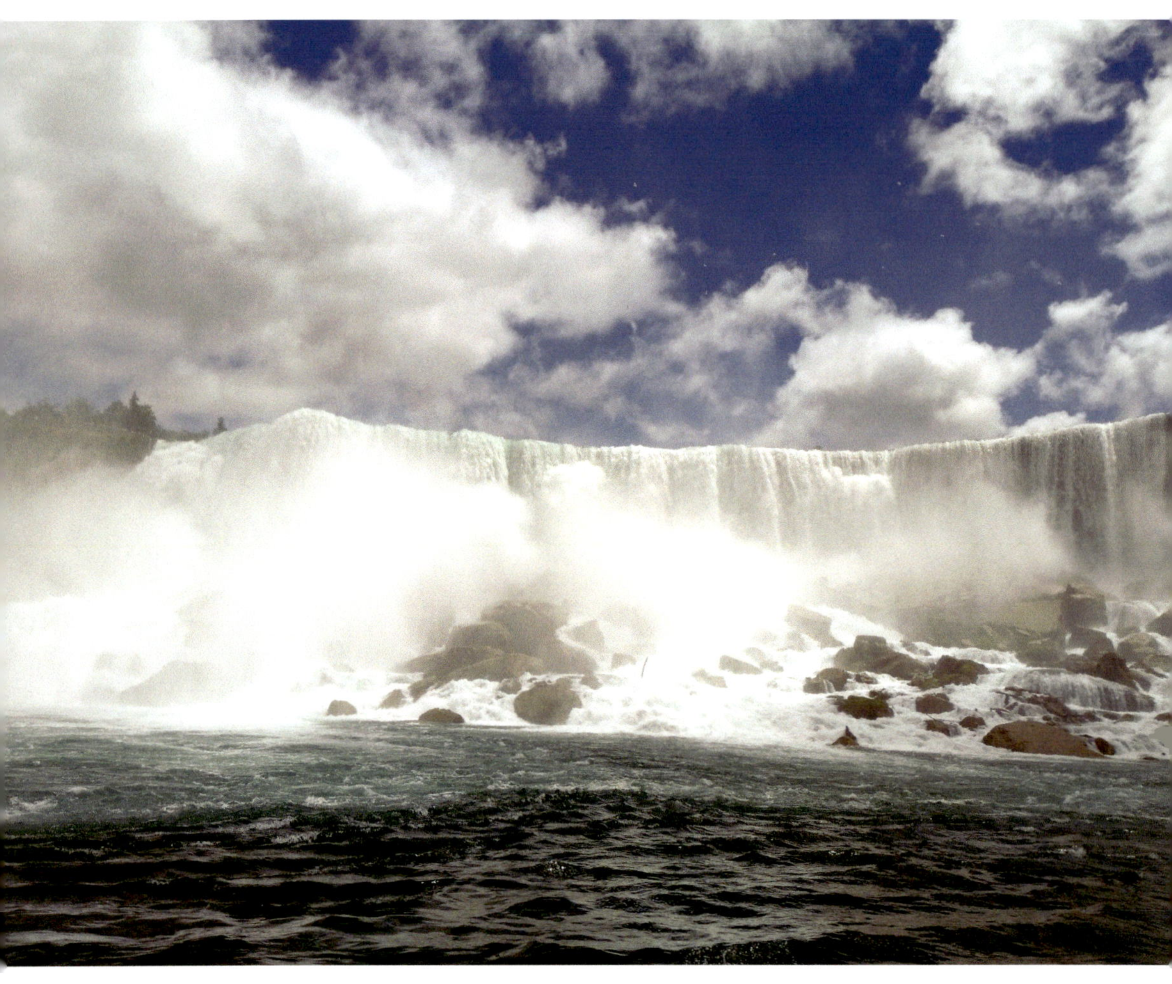

The full majesty of the falls is best viewed from the river itself, on one of the tour boats that offers close-up views.

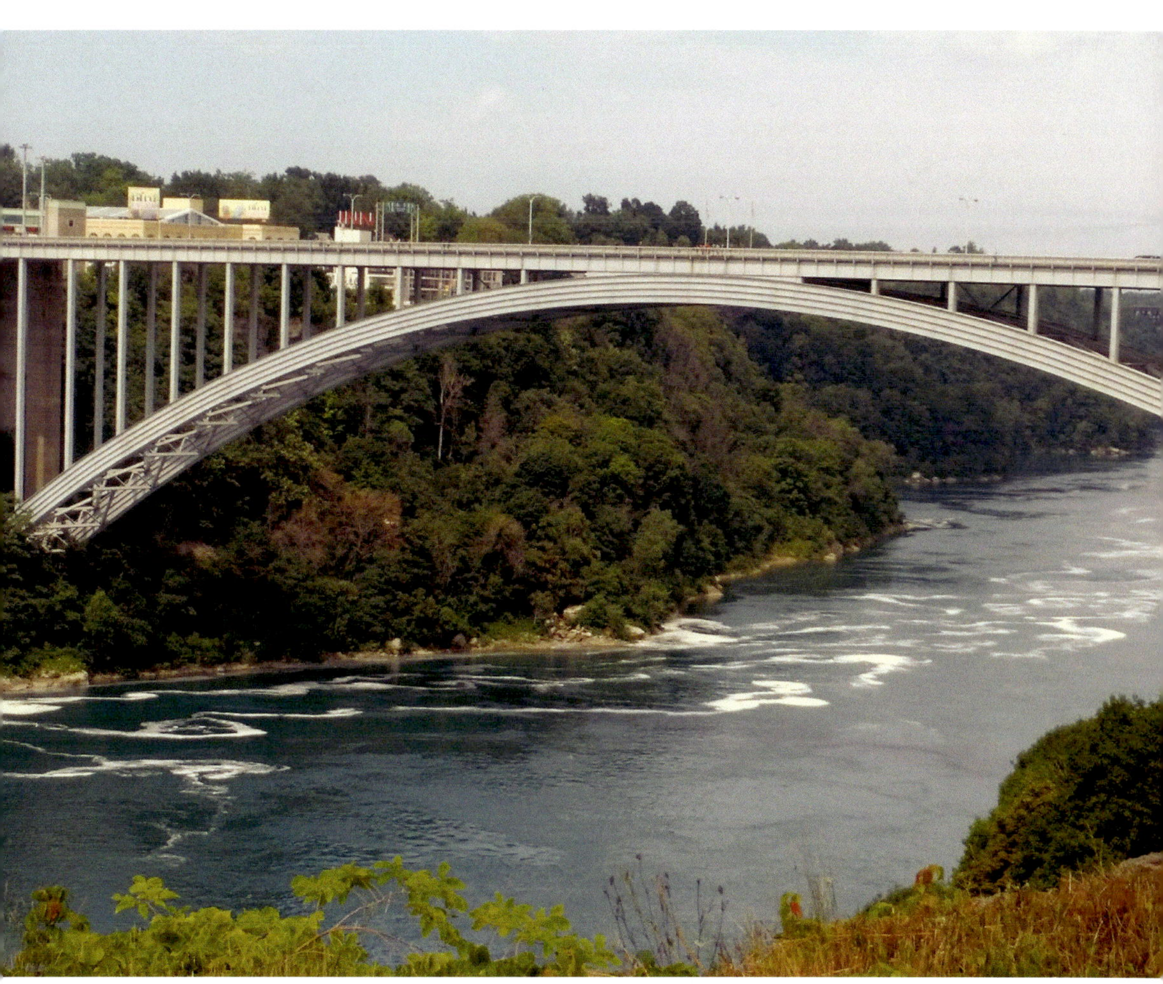

The Rainbow Bridge as seen from the Canadian side in 2018.

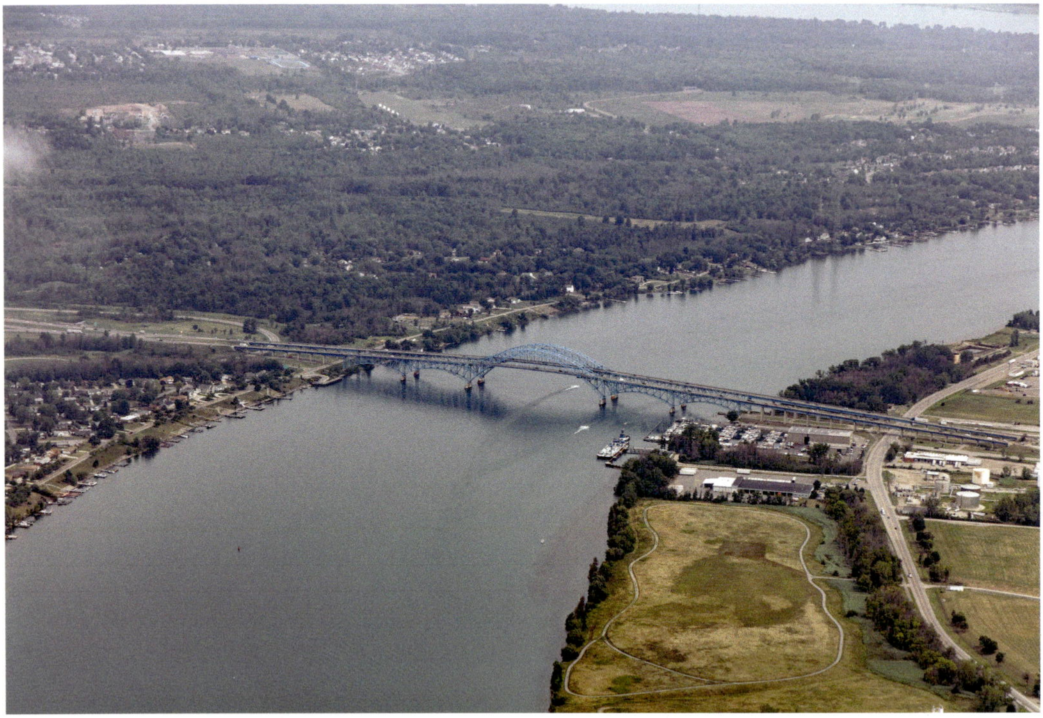

Above and below: Two aerial views of the twin spans of the South Grand Island bridge taken by Carol Highsmith, 2018. (*Library of Congress*)

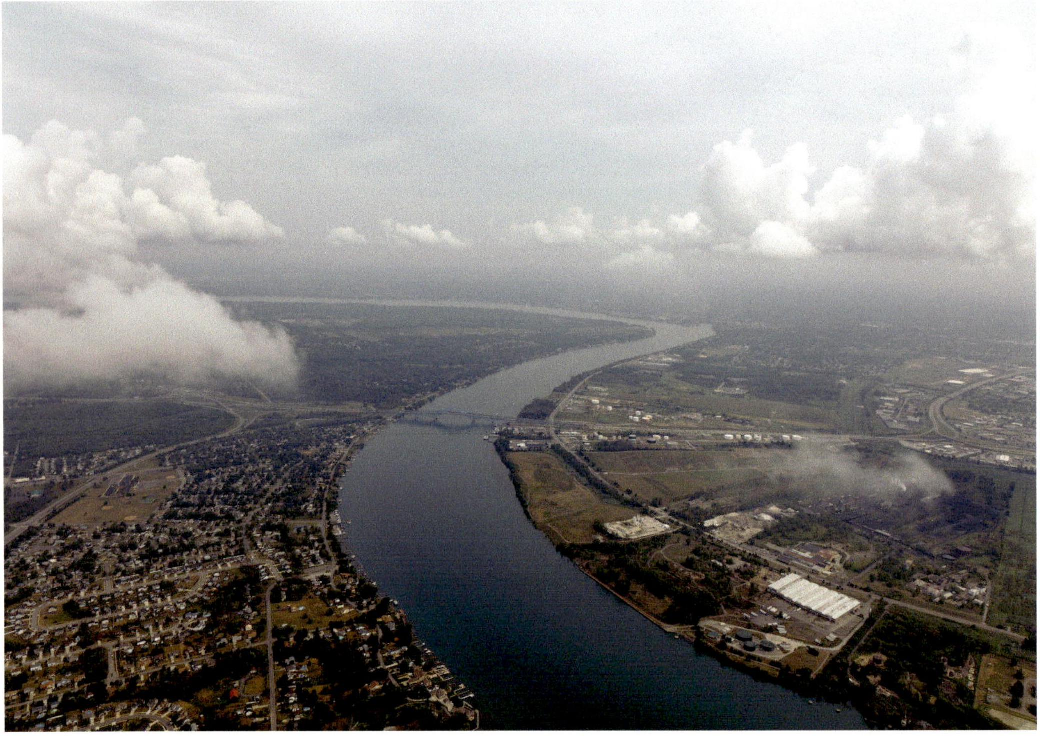

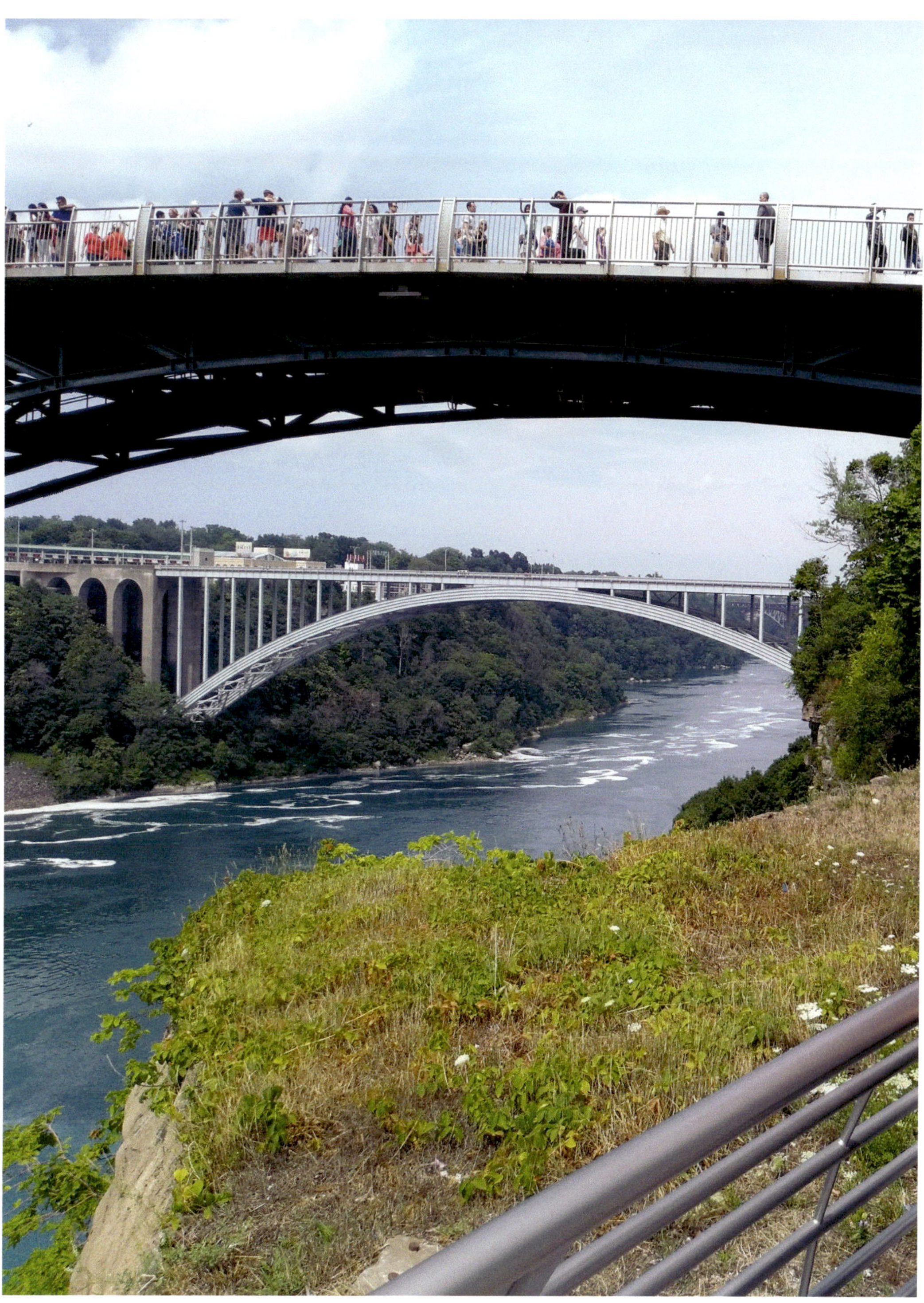

The 282-foot-high Niagara Falls Observation Tower, seen here in 2018 with the Rainbow Bridge in the background, was built in 1961 on the American side of the river. It offers views of the falls for millions of visitors annually. It looks like an unfinished bridge, but it is actually just the way it was intended.